AS UNIT 3 | A2 UNIT 3

STUDENT GUIDE

CCEA

Geography

Fieldwork skills and techniques in geography

Decision-making in geography

Tim Manson

HODDER
EDUCATION
AN HACHETTE UK COMPANY

Hodder Education, an Hachette UK company, Blenheim Court, George Street, Banbury, Oxfordshire OX16 5BH

Orders

Bookpoint Ltd, 130 Park Drive, Milton Park, Abingdon, Oxfordshire OX14 4SB

tel: 01235 827827

fax: 01235 400401

e-mail: education@bookpoint.co.uk

Lines are open 9.00 a.m.–5.00 p.m., Monday to Saturday, with a 24-hour message answering service. You can also order through the Hodder Education website: www.hoddereducation.co.uk

© Tim Manson 2016

ISBN 978-1-4718-6411-7

First printed 2016

Impression number 5 4 3 2 1

Year 2021 2020 2019 2018 2017 2016

This guide has been written specifically to support students preparing for the CCEA AS and A-level geography examinations. The content has been neither approved nor endorsed by CCEA and remains the sole responsibility of the author.

Cover photo: Alistair Hamill. Other photos: p.17 Michael Raw; p.10, pp.64–67 Tim Manson. p.64, Figure 1 © Crown Copyright Ordnance Survey (Account No. 100036470). p.69, Resource E reproduced by permission of www.breakingnews.ie; p. 70 Resource F reproduced by permission of Tourism Northern Ireland.

Typeset by Integra Software Services Pvt. Ltd, Pondicherry, India

Printed in Italy

Hachette UK's policy is to use papers that are natural, renewable and recyclable products and made from wood grown in sustainable forests. The logging and manufacturing processes are expected to conform to the environmental regulations of the country of origin.

Contents

Content Guidance

AS Unit 3

The geographical fieldwork process • Planning for a fieldwork study • Sampling techniques • Fieldwork safety • Data collection techniques • Data organisation and presentation • Data analysis and interpretation • Drawing conclusions and evaluation • Preparing the fieldwork report and table

Data collection • Data processing • Methods of statistical analysis

A2 Unit 3

Key attributes of the decision-making process • Using the resources • Tips and techniques • Advanced statistical analysis (for A2 only)

Questions & Answers

AS Unit 3 • A2 Unit 3 • Information for both papers

AS Unit 3

A2 Unit 3

■ Getting the most from this book

Exam tips

Advice on key points in the text to help you learn and recall content, avoid pitfalls, and polish your exam technique in order to boost your grade.

Knowledge check

Rapid-fire questions throughout the Content Guidance section to check your understanding.

Knowledge check answers

1 Turn to the back of the book for the Knowledge check answers.

Summaries

■ Each core topic is rounded off by a bullet-list summary for quick-check reference of what you need to know.

Sample student answers

Practise the questions, then look at the student answers that follow.

Exam-style questions

Commentary on the questions

Tips on what you need to do to gain full marks, indicated by the icon **e**

Questions & Answers

e 3/3 marks awarded. This makes reference to three different aspects of safety equipment that were used for this particular fieldwork. It discusses both the role of the equipment and the importance of this within the context of their visit to the sand dune.

(b) Describe and explain **one** sampling method that was used within your fieldwork. [6 marks]

e You can select from a range of different sampling methods, but most relate to one of either random, systematic, pragmatic or stratified sampling. 3 marks are for a general description of the sampling method and the other 3 marks are for an explanation of how the sampling method works in the context of this particular fieldwork.

(b) The aim of our fieldwork was to visit a total of 15 sites across the sand dune ecosystem at Magilligan Point. We were trying to see how the infiltration rate (how quickly 200 ml of water soaked into the soil) changes as you move away from the sea.

Our sampling method was to use systematic sampling to take measurements at sites along a line stretching back from the sea through the sand dune system. We took measurements at 15-m intervals along this line. We did this as other sampling techniques would not have been right for this coursework. We wanted to see how the infiltration changed through the sand dune and not just at random places. Pragmatic sampling might have worked too as the problem with systematic sampling was that sometimes the sites were not showing the changes that actually happened in the sand dune.

e 5/6 marks awarded. This is a good description of how the sampling technique was used within the context of this fieldwork. There is good depth about how systematic sampling was used to look at the changes across the sand dune system. The answer then goes on to explain why this sampling method was appropriate and even mentions how another sampling method might have been used. However, this explanation could have been more focused.

(c) (i) Choose **one** of the following statistical techniques that could be used to analyse some of your fieldwork data. Your chosen technique must fit in with the aim of your fieldwork.
  ■ Spearman's rank correlation
  ■ Nearest neighbour analysis
  ■ Mean, median, mode and range
In the space provided, complete your chosen statistical analysis and show all your calculations clearly. If relevant, comment on the level of statistical significance. (Significance graphs and formulae are provided — see pages 12 and 30.) [7 marks]

e The statistical technique depends on the chosen fieldwork, but it needs to be relevant to the aim/hypothesis of the investigation.

Commentary on sample student answers

Read the comments (preceded by the icon **e**) showing how many marks each answer would be awarded in the exam and exactly where marks are gained or lost.

■ About this book

Much of the knowledge and understanding needed for AS geography builds on what you have learned for GCSE geography, but with an added focus on geographical skills and techniques, and concepts. This guide offers advice for the effective revision of **Unit AS 3: Fieldwork skills and techniques in geography** which all students need to complete. In addition, this guide offers guidance on **Unit A2 3: Decision-making in geography**.

The AS 3 external exam paper tests your knowledge and application of geographical skills and techniques and lasts 1 hour. Students answer two structured questions: Question 1 on fieldwork skills and Question 2 on geographical techniques. The unit makes up 20% of the AS award or 8% of the final A-level qualification.

The A2 3 external exam paper allows students to practise decision-making skills through the application of a real world scenario in the form of a case study. The exam paper lasts 1 hour 30 minutes and the unit makes up 12% of the final A-level qualification.

To be successful in these units you have to understand:

■ the key ideas of the content
■ the nature of the assessment material — by reviewing and practising sample structured questions
■ how to achieve a high level of performance in the exam

This guide has two sections:

Content Guidance — this summarises some of the key information that you need to know to be able to answer the examination questions with a high degree of accuracy and depth. Students will also benefit from noting the **Exam tips,** which will provide further help in determining how to learn key aspects of the course. **Knowledge check** questions are designed to help learners to check their depth of knowledge — why not get someone else to ask you these?

Questions & Answers — this includes some sample questions similar in style to those you might expect in the exam. There are some sample student responses to these questions as well as detailed analysis, which will give further guidance in relation to what exam markers are looking for to award top marks.

The best way to use this book is to read through the relevant topic area first before practising the questions. Only refer to the answers and commentary after you have attempted the questions.

Content Guidance

AS Unit 3

■ Fieldwork skills

The first question on the AS Unit 3 paper asks you to make reference to some fieldwork/data collection that you have taken part in. Due to the diversity of fieldwork opportunities at AS level, it is impossible for this guide to refer to every type of geography enquiry. Instead, it will cover one common example from the specification. However, the process and recommendations can be easily transferred to any piece of work.

When tackling any geographical enquiry, the first task is to identify the particular question and issues that are to be investigated.

The geographical fieldwork process

Most pieces of geography enquiry follow a similar investigation path:
- Title/aims/hypotheses
- Planning
- Data collection
- Data organisation (tabulation and presentation)
- Data analysis and data interpretation
- Drawing conclusions and evaluation

Exam tip

Make sure you understand the sequence of this investigation path.

Planning for a fieldwork study

Many geography teachers are keen to involve students in the collection of primary data in order to develop their geographical skills.

Location selection

When planning a field trip, it is important to consider some key questions regarding the best location for your study:
- Is the location accessible?
- Is it easy to get to?
- Is it appropriate for a group of students to visit?
- Do we need permission to access the site?
- How much will it cost to get there or to get in?
- How much time will it take us to get there?
- Will it be safe?
- Could we damage the environment by being there?
- Is this the best example of this type of environment/feature?

Exam tip

Apply these questions to your field trip and address any potential planning issues.

Risk assessment

No geography field trip can be undertaken in a hazard-free environment. Both students and teachers need to consider any risks before going into the field so that appropriate measures can be taken to ensure safety. For example, one risk in an area of sand dunes might be rabbit holes and undulating surfaces, which can be managed by advising students to wear sturdy shoes and to watch where they are walking.

Contingency plans need to be made to manage any risk. For school trips, a comprehensive risk assessment must be submitted to the Principal to demonstrate how any potential situation might be handled.

Exam tip

Ask your teacher if you can see the risk assessment they had to submit before taking you on the field trip.

Pilot study

It might be important to conduct a pre-study site visit to make sure that the location is appropriate for the needs of the enquiry. You might need to check the accessibility, conduct a risk assessment and make sure that the results taken will allow the aims/hypotheses to be addressed. Equipment could be tested to make sure that it is appropriate for the study. If using a questionnaire survey, it might be good practice to test the questions a few times to make sure that they read well and make sense to the people you are testing.

Sampling techniques

The purpose of fieldwork is to enable you to collect your own primary data, which might support any additional material gained from secondary sources.

- Primary sources/data comprise any new information that you have collected in the field. This might be done through observation or through measurement.
- Secondary sources/data comprise any new information that has been obtained from existing sources, such as maps, Geographical Information Systems (GIS), photographs or census data. Often in geography investigations you will need to use both primary and secondary information.

Knowledge check 1

What are some of the strengths and weaknesses of using primary data?

During field visits, sampling is needed because time restrictions make it impossible to study and take measurements from an entire area. Therefore, decisions have to be taken as to what is the most appropriate method of choosing which locations and aspects of the study are recorded.

The main sampling techniques are random, systematic, stratified (point, line and quadrat) and pragmatic.

Random sampling is when a random number table or random number generator app is used to give, for example, the sequence of people to ask, or houses to call in a street when doing a survey.

Systematic sampling is when samples are taken using a pre-determined interval. For example, questionnaires might be completed for every 5th or 10th person who walks past, or soil studies might be taken every 5 m or 10 m along a survey line/transect.

Stratified sampling is a useful way of sampling when there are clear sub-groups within the dataset. For example, if conducting a sample of 70 questionnaires in a school population of 700 students, you might break the sample down so that those

Exam tip

Make sure that you know the differences between these sampling techniques. Questions commonly ask about the integrity of a sampling technique and how it was applied to your fieldwork.

questionnaires allocated to a year group represent the size of that year group within the school. So, for example, if there were 100 students in the sixth form they would get 10 questionnaires.

Point sampling is when individual points are used within the investigation. For example, specific, accessible sites might be chosen in a river study.

Line or transect sampling is when a line is drawn on a map within an area and all data are collected along this line.

Quadrat sampling is when a piece of equipment called a quadrat is used to measure the amount of vegetation/type of vegetation or amount of ground coverage within an area. The quadrat is usually a square metal frame (the most common is 50 cm × 50 cm).

Pragmatic sampling is when decisions are taken to visit sites that are safe/accessible or which might demonstrate typical characteristics. Although this approach often allows for a simple fieldwork experience, it often introduces a huge amount of bias into the sampling technique.

Fieldwork safety

Safety during fieldwork is very important and must be considered when preparing to go into the field (this will form part of any risk assessment strategy).

Weather conditions

Wet and freezing conditions can cause problems on field trips, while strong winds can be an issue when using fieldwork equipment. Wet clothes are uncomfortable and can cause rapid heat loss from a body, leading to hypothermia.

Strategy: Check the weather forecast and take appropriate clothing for the weather conditions and perhaps a change of clothes, a first aid kit and a thermal blanket.

Injuries

Many physical environments — for example, the coast, uplands, rivers and forests — can be very dangerous places if care is not taken. Beaches, sand dunes and riverbeds have uneven surfaces. Slopes can be steep, uneven or have loose material that could cause a fall.

Strategy: Wear appropriate shoes/boots, avoid running, watch where you are going, carry a first aid kit with dressings and perhaps a bivi bag to carry an injured student.

Fieldwork equipment can be dangerous. For example, ranging poles, clinometers and metre sticks can be dangerous if carried in the wrong way. Some infiltration rings might have sharp edges, while safety ropes can cause 'rope burn'.

Strategy: Carry ranging poles with the spike down. All equipment should be checked to make sure there are no sharp edges or rust. Ropes should be rolled up carefully and carried responsibly.

Exam tip

Consider the number of sites/questionnaires that you need for your sample. Often this is linked to the statistical technique that you select. For example, if you use Spearman's rank correlation you will need to visit at least 15 sites.

Safety in urban areas

When working in urban areas, risks still exist in relation to traffic and other people. You should remain vigilant, ensuring that you do not get separated from your group.

Strategy: Use the green cross code when crossing roads, stay within groups, establish designated check-in times and places and have emergency contact numbers available.

Data collection techniques

The data collection techniques that you choose depend on the title and hypotheses that you have selected to investigate. Here (and in later sections) we will look at one example of physical fieldwork, but you might want to consider how your data collection/fieldwork experience prepared you for this section of the exam paper.

Fieldwork title: An investigation into seral succession of the psammosere at Magilligan Point

Hypotheses:

1 Infiltration time increases with distance travelled from the sea.
2 Vegetation becomes more complex with distance travelled from the sea.
3 Ground cover increases with distance travelled from the sea.

Data collection methodology

In order to measure the changes to the sand dune system in relation to the distance from the sea, a systematic/transect sampling technique was used to collect results. Students started at the edge of the sea and had to move along a straight line, taking measurements at 15 sites at 15-metre intervals. This sample size was chosen as it is the minimum number needed to complete an accurate Spearman's rank correlation.

At each site, infiltration, vegetation characteristics and ground cover were measured.

Infiltration: An infiltrometer (infiltration ring — Figure 1), measuring cylinder, water and stopwatch were used to measure the rate of water infiltration at each site. The vegetation was cleared from the test area and 200 ml of water was poured into the infiltrometer, which had been pressed into the soil. The time taken for the water to clear into the soil was recorded using the stopwatch. This was repeated three times at each site to get an average time.

Vegetation characteristics and ground cover: A 50 cm × 50 cm quadrat and species identification list were used to identify the type of vegetation at each site. The quadrat was 'tossed' randomly near the site and any vegetation type and the percentage of different types of ground coverage were noted on a results table.

Exam tip

Make sure that you have detailed knowledge of any equipment you use and can explain how it produced your results. Did you have any challenges in collecting your data?

Figure 1 Using an infiltration ring on a sand dune

Data organisation and presentation

The first step in sorting your data is to create a table of results. This should be brought into the exam as part of your prepared material.

Table 1 Results table for investigation into seral succession at Magilligan Point

Site (distance from the sea in metres)	Time (for 200 ml of water to infiltrate, in seconds)	Number of different species of vegetation	Vegetation cover (%)
1 (0)	10.1	0	0
2 (15)	32.3	0	0
3 (30)	62.5	1	20
4 (45)	95.5	2	35
5 (60)	120	4	50
6 (75)	133	2	55
7 (90)	62	1	65
8 (105)	156	5	80
9 (120)	132	6	90
10 (135)	171.4	6	95
11 (150)	235	7	95
12 (165)	245	5	95
13 (180)	266	7	90
14 (195)	256	8	95
15 (210)	262.5	8	100

Exam tip

Make sure that you prepare your data table carefully and follow the instructions issued by the awarding body. The more organised your table is, the quicker you can answer the exam questions.

Some of the questions on the exam paper expect you to use the fieldwork table that you bring into the exam either to produce a **graph** or to apply a **statistical technique**. Further details on the different statistical techniques are found later in this book.

Using just simple methods of statistical analysis (e.g. mean, median, mode and range) can cause more difficulties when trying to analyse and interpret the results later in the exam paper. Therefore, it might help to choose one hypothesis to which you can apply the Spearman's rank correlation.

For hypothesis 1: Infiltration time increases with distance travelled from the sea — the following Spearman's rank table can be drawn from the results.

Table 2 Spearman's rank correlation table for fieldwork example

Site (distance from the sea in metres)	Rank	Time (for 200 ml of water to infiltrate, in seconds)	Rank	Difference in ranks, d	d^2
1 (0)	15	10.1	15	0	0
2 (15)	14	32.3	14	0	0
3 (30)	13	62.5	12	1	1
4 (45)	12	95.5	11	1	1
5 (60)	11	120	10	1	1
6 (75)	10	133	8	2	4
7 (90)	9	62	13	−4	16
8 (105)	8	156	7	1	1
9 (120)	7	132	9	−2	4
10 (135)	6	171.4	6	0	0
11 (150)	5	235	5	0	0
12 (165)	4	245	4	0	0
13 (180)	3	266	1	2	4
14 (195)	2	256	3	−1	1
15 (210)	1	262.5	2	−1	1

$\Sigma d^2 = 34$

The Spearman's rank formula is applied as follows:

$$r_s = 1 - \left(\frac{6\Sigma d^2}{n^3 - n}\right)$$

$$r_s = 1 - \left(\frac{6 \times 34}{15^3 - 15}\right)$$

$$r_s = 1 - \left(\frac{204}{3,360}\right)$$

$$r_s = 1 - 0.06$$

$$r_s = 0.94$$

An r_s result of 0.94 shows that there is a very strong positive correlation or relationship between the two variables (time it took 200 ml to infiltrate through the soil and distance from the sea).

Exam tip

You need to have practised one graphical representation and one statistical analysis. For this example I would suggest that you prepare Spearman's rank analysis and a scattergraph.

Exam tip

An explanation of how to work through the use of the Spearman's rank technique can be found later in this book.

Exam tip

Don't forget that this table uses information from the data table (Table 1) and cannot be pre-prepared.

The graph and table in Figure 2 will usually be provided to help you to determine the significance of your result. In this case a critical value of 0.94 is within the 99.9% significant area. This means that the result is very significant and the relationship between the two variables can be commented on.

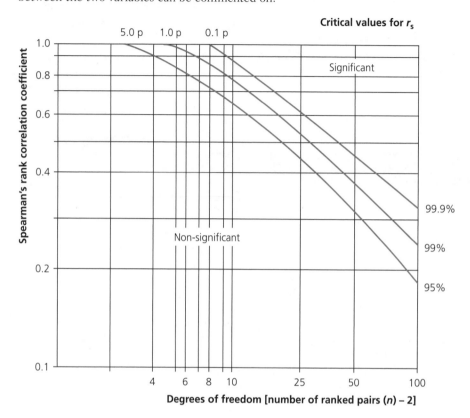

Degrees of freedom	Significance level	
	0.05 (5%)	0.01 (1%)
8	0.72	0.84
9	0.68	0.80
10	0.64	0.77
11	0.60	0.74
12	0.57	0.71
13	0.54	0.69
14	0.52	0.67
15	0.50	0.65
20	0.47	0.59

Figure 2 Spearman's rank correlation significance graph and table

Other questions in relation to data presentation might ask you to use data from the table to draw a graph that is relevant to the aim of the fieldwork. Make sure that you draw a graph that allows the opportunity for a full analysis in later questions. For hypothesis 1: Infiltration time increases with distance travelled from the sea, a scattergraph (Figure 3) or a line graph is appropriate.

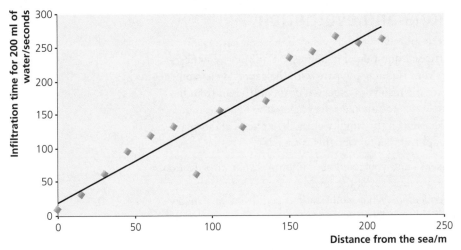

Figure 3 Scattergraph to show time taken for water to infiltrate through a sand dune against the distance from the sea

Care must be taken when drawing the graph as marks are awarded for the accuracy of your presentation. In previous exam series, 7 marks were awarded for a graph, with 1 mark for a specific and accurate title, 2 marks for the use of mathematical conventions (labelled axes, use of a key or scaling of the graph), 3 marks for accuracy (the precise plotting of values) and 1 mark for the method in selecting an appropriate graphical technique in relation to the aim and data table.

Data analysis and interpretation

After drawing your graph or using a statistical technique on your data, you then need to analyse or interpret your findings with reference to your stated aim.

Data analysis

You might be asked to *describe* the data, so you should practise writing about the *patterns and trends* that you notice in the results. Use figures from the graph/table/statistical technique to add weight to your description. Note the highs/lows and any averages that you have calculated. Describe any relationships on the graph — such as positive or negative correlations — and comment on the significance of the result. Describe any unusual results.

Data interpretation

You might be asked to *explain* or *interpret* the data, so practise making reference to the results shown in the graph/statistical technique and then try to explain the reasons behind your results. What are the factors that have controlled or created this situation? Do your results help you to prove or disprove the hypothesis that you have been testing? How does this fit within general geographical theory? Is this what you expected? Are there results that go against what you expect?

Often you are asked to reflect on the aims and/or hypothesis that you had stated on your fieldwork report. Practise referring back to the aim/hypothesis and show how using this technique has helped you to prove or disprove the hypothesis.

Exam tip

The accuracy of your statistical analysis measure or graph is extremely important — many students lose easy marks here by rushing. Make sure you have a sharp pencil!

Exam tip

Think carefully about why you have chosen your particular method of statistical analysis or graphical presentation and make sure that you can explain why it is suitable for your piece of fieldwork.

Exam tip

Practise analysing and interpreting your graph, and statistical analysis. These are common questions on the exam paper.

Drawing conclusions and evaluation

A **conclusion** is a summary of the information and evidence that you have been considering as you address a particular question or hypothesis. Can your hypothesis be accepted or rejected? How do your findings compare with the general geographical theory associated in this area? Are the results as expected or very different from the expected? Why might that be the case? You should refer back to the hypothesis and show which data support your decision on the validity of the hypothesis and explain (with reference to relevant geographical theory) why this is the case.

An **evaluation** allows you to discuss how you could have improved your investigation. What are the main strengths and weaknesses of the investigation? What are the limitations of the evidence you collected? What modifications could you have made to improve the accuracy of your fieldwork? How could your investigation have been further extended?

How could you have improved your data collection methods?

Was your methodology accurate enough to produce results that you could trust? How could you have gone further to make sure that your results were accurate? What additional equipment might you have used to get more accurate results?

How could you have improved your sampling technique?

Was it fit for purpose? Would a different sampling technique have allowed more accurate results? Did you visit too many/too few sites?

How could you have improved your conclusions?

Are your conclusions reliable and accurate? Were the title/aim/hypotheses that you selected appropriate for what you wanted to study? Was the location of your study appropriate?

Preparing the fieldwork report and table

Remember that you are expected to submit a fieldwork report and data table to your teacher or examinations officer before the examination. The focus for this piece of fieldwork should be an issue, hypothesis or question that arises from the AS Unit 1 or 2. Information used to prepare for this question must include both primary and secondary sources.

The **fieldwork report** or a summary statement (a short report of around 100 words) should include:

- a general title for the fieldwork
- a brief outline of the spatial location of the study (including a location map, if desired)
- a brief statement of the aims/purpose/issues that provide the theoretical context for the personal investigation element of the fieldwork (this can be key questions/hypotheses)

The report should not address any other aspects of the investigation.

The **data table** (an attached table of data) should include:

- a specific title
- data collected for all variables relevant to the proposed aim/purpose of the study in the report

Exam tip

Hand your finished written report and data table to your geography teacher a few days before the examination and make sure that you fill in and sign the declaration sheet.

- primary and secondary data essential to the aim
- quantitative data (numbers), essential to allow graphical representation and statistical analysis
- adherence to normal conventions (all variables stated clearly and precise units of measurement)
- raw data (averages or other statistical calculations should not be included)

Summary

- In preparation for the fieldwork skills question on the paper (Question 1) you need to bring a fieldwork report and a data table into the exam with you.
- Make sure that you understand the usual order of investigation within the fieldwork process.
- Geography field trips require careful planning, risk assessment and measures to ensure health and safety at all times.
- A variety of data collection techniques can be used to observe and measure geographical data.
- Most fieldwork will involve collection of primary data.

- You should consider the best sampling technique to ensure the integrity of your recorded data.
- A statistical technique such as Spearman's rank analysis should be practised and applied to an aspect of the data table.
- A graphical technique such as a scattergraph should be practised and applied to an aspect of the data table.
- Practise referring to your hypothesis/aim to prepare for questions on data analysis, data interpretation, conclusions and evaluation.

Geographical techniques

The second question on the AS Unit 3 examination paper will test your knowledge and application of a series of geographical skills and techniques — from either the physical or human side of the course. You will be expected to respond to a range of quantitative and qualitative questions. You are expected to be able to use these skills in the other examination papers as well. The diversity of data collection and processing opportunities and stimulus material in A-level geography means that it is impossible in this short guide to refer to every possible map or graph type available.

You are expected to have knowledge of:
- **data collection:**
  - using equipment, surveys and questionnaires
  - analysis and interpretation of maps at a range of scales (public maps and Ordnance Survey), photographs, satellite images, surface pressure or synoptic charts, remotely sensed images and data from secondary sources including geographical information systems (GIS)
- **data processing:**
  - map skills — map distributions, densities and flows using dots, flow lines, choropleth shading and isolines
  - sketches — drawing annotated sketch maps
  - graphical skills — constructing, analysing and interpreting scattergraphs, line graphs, bar graphs, pie charts, proportional graphs and triangular graphs (including titles, keys, scales, frames and direction arrows)
  - sampling methods — sampling, including random, systematic, stratified (point, line and quadrat) and pragmatic

- statistical analysis — using mean, median, mode and range; Spearman's rank correlation and nearest neighbour analysis
- **embedded skills** — although not specified in detail, you are expected to be able to use, and refer to use of, GIS and the internet. You are encouraged to use ICT for collecting, sorting, recording and presenting geographical information. These skills are not covered in this guide

Data collection

Much of the information on fieldwork/data collection can be found earlier in this book. However, some of the data collection methods could be referenced in questions in this part of the examination paper.

Using surveys and questionnaires

Surveys and questionnaires are used in many different ways in human geography. Data comprise the information that you collect in order to address the aims or hypotheses within any investigation. Questionnaires can be a good method of collecting information and opinions that people have. You need to think carefully about the sample size (what is the appropriate number of questionnaires to ask in relation to a study?) and also consider the number of questions and length of the questionnaire — it always takes longer than you expect to administer a questionnaire.

- Primary sources/data refer to new information that you have collected in the field. This might be done through observation or through measurement.
- Secondary sources/data refer to information that has been obtained from any other source.

You will often need to use both primary and secondary information.

The main sampling techniques are random, systematic, stratified (point, line and quadrat) and pragmatic. Sampling techniques are needed to reduce bias in the investigation. Some of the techniques introduce more bias into the study than others. (See below for more information on sampling methods.)

> **Exam tip**
>
> Make sure that you can clearly demonstrate the difference between primary and secondary sources.

Maps, charts and GIS

There is a wide range of maps which could be used as a stimulus for exam questions in this section. Students will come across a variety of examples of this as they work through the Unit 1 and Unit 2 course. Examples of the different uses of satellite images and surface pressure charts can be found in the Unit 1 Student Guide.

Questions using GIS are becoming increasingly popular. Students should practise analysing GIS data in relation to census information using the Northern Ireland Neighbourhood Information Service (NINIS) information on the internet. Further GIS data can be found in relation to rivers and flooding information with the Northern Ireland Rivers Agency.

Data processing

Geography exams require candidates to be able to draw a range of maps, charts and diagrams as part of their answers. You need to make sure that you understand how to construct, analyse and interpret a range of images. You should also make sure that you can read and interpret Ordnance Survey maps (at a range of scales).

Map skills

Photos

Questions might ask you to label or annotate a photograph such as Figure 4.

Figure 4 A river channel

Each label should be a word or sentence that describes or identifies a feature shown in the photograph.

Satellite images

Satellite images might be used to allow students to analyse and interpret geographical patterns within either a physical or human environment. You need to use each resource carefully and should look for patterns on the map and describe these geographically.

Dot distribution maps

Dot maps can be used to show a distribution pattern within an area. Each dot will represent a specific value and the number of dots in an area will indicate the density and distribution of this within the population.

Figure 20 on page 31 is an example of a dot distribution map. It shows the distribution of Met Office automatic weather stations across Northern Ireland.

Flow line maps

Flow lines can be used on a map to show the amount/volume and the direction of a movement from one place to another (Figure 5). Usually, the width of the flow line will indicate the amount of movement in this direction (used in conjunction with a scale key). These maps can be useful to illustrate, for example, migration streams or flows of tourists from one place to another.

Exam tip

Make sure that you annotate geographical features as fully as possible.

Knowledge check 2

Draw an annotated sketch of the river channel in Figure 4 and any other typical river features noted in the picture.

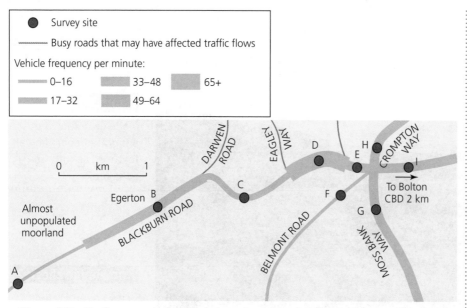

Figure 5 Traffic flows on the main roads to the north of Bolton, Greater Manchester

Knowledge check 3

Use Figure 5 to answer the following:

(a) On which section of road is the highest amount of traffic?

(b) Name two roads where there is very little traffic.

Choropleth maps

Choropleth maps (sometimes called 'area-shaded' maps) are shaded according to the density of a value in an area (Figure 6). Usually, any data used in the construction of a choropleth will be grouped into around 5–7 different categories. These will then be plotted and shaded appropriately on the map.

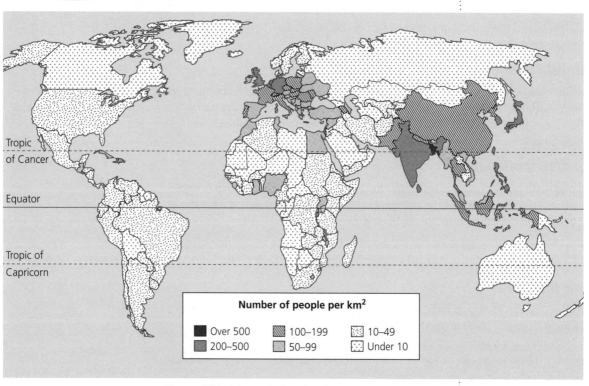

Figure 6 World population density by country, 1996

Some geographers argue that these maps oversimplify the data because they can suggest that a large area has the same value. For example, in Figure 6, all of China is shown as having between 100 and 199 people per km^2. However, this does not reflect the fact that some parts of China are much more densely populated and some are sparsely populated. Maps like this often suggest that there are huge changes at boundaries (e.g. the UK and Ireland) whereas this might not be the case.

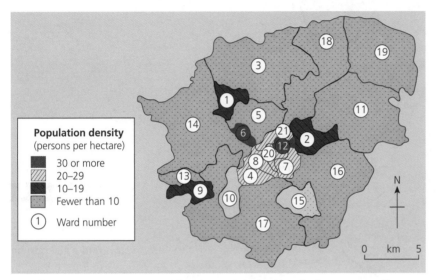

Figure 7 A choropleth map showing population density in a metropolitan borough in northern England, 2001

Isoline maps

An isoline is a line that joins places on a map that have the same value (Figure 8). Some of the most commonly used isolines are isobars (where areas of the same high pressure on a weather map are joined), isotherms (where areas of the same temperature are joined) and contours (where areas with the same height on an OS map are joined).

Sketches

Field sketches

These are similar to the annotated photos mentioned above. The key is to use labels and annotations to describe and explain some of the features and processes shown in each resource.

Knowledge check 4

Use Figure 7 to answer the following:
(a) Which two wards have the highest population density?
(b) How many wards have population density between 10 and 19 people per hectare?

Exam tip

Choropleth maps are common in exam papers as they can challenge the student to think carefully. Think about how you might go about constructing a choropleth map. What are the advantages of displaying your data in a format like this?

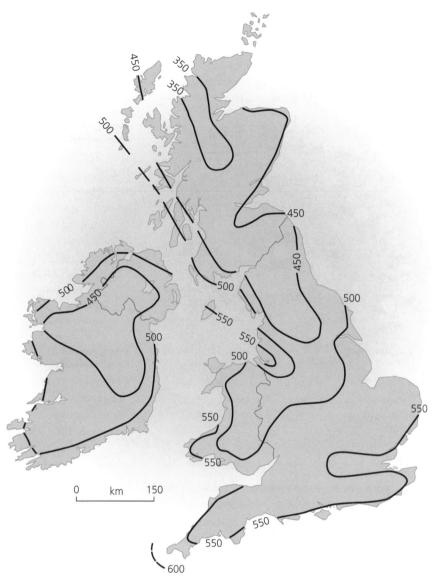

Figure 8 Potential evapotranspiration in the British Isles (mm)

Knowledge check 5

Use Figure 8 to answer the following:
(a) Add a 400 mm line to Figure 8.
(b) Describe the pattern of potential evapotranspiration in the UK (don't forget to describe the location and use figures to support your answer).

Annotated sketch map

Sometimes, examiners might ask students to draw a sketch map from a grid square or series of grid squares on a map. You are usually given an OS map and asked to copy some of the main features onto an enlarged, blank grid square. When drawing a sketch map remember to:

■ add a title
■ use as many labels as possible to describe what you see on the map
■ add a scale
■ add a key to show any features you have added
■ show any contours
■ add a north arrow
■ mark any grid squares

Exam tip

Always draw maps and graphs neatly and accurately so that you maximise your mark in the exam.

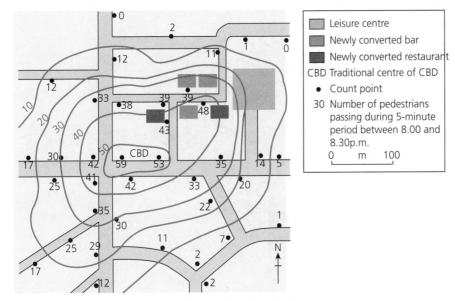

Figure 9 Isoline map showing pedestrian densities in a town centre

Graphical skills

Bar graphs

Bar graphs (Figure 10) are common in geography and can be presented using a number of different formats.

A common misconception is that bar graphs and histograms are the same. A histogram is usually used to display continuous data and has blocks connected to each other. It also shows numbers (or frequencies) of a whole sample and should be all shaded in the same colour.

Bar graphs are usually used to display discrete (non-continuous) data, with the blocks separated by a space. As each bar represents different, discrete data it is shaded in a different colour.

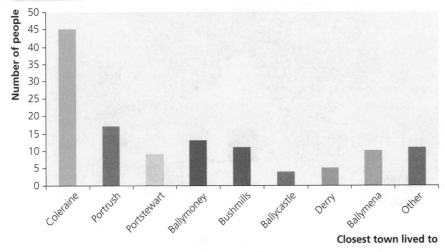

Figure 10 A bar graph to show the town nearest to where people live in a population survey in Coleraine town centre

Knowledge check 8

Using Figure 11, describe the main areas of net migration increase and loss between 1990 and 1995.

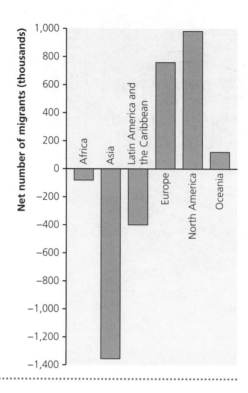

Figure 11 A gain-loss bar chart showing annual net migration totals in the world's major areas, 1990–95

Knowledge check 9

From Figure 12, describe the global distribution of the squatter population (in 2000).

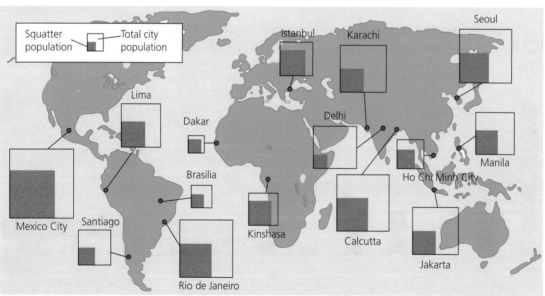

Figure 12 Map with proportional squares to show the size of the squatter population in selected world cities, 2000

Line graphs

Line graphs (Figure 13) can be effective for displaying continuous data or measurements and can be used to show changes over time.

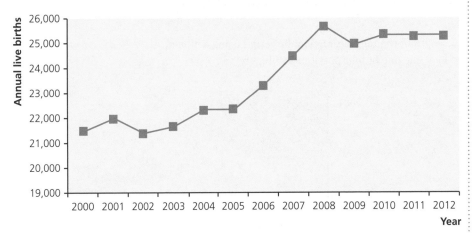

Figure 13 A line graph showing live births in Northern Ireland, 2000–2012

> **Exam tip**
>
> Often students think that bar graphs and line graphs can be used for any presentation of data. This is not the case. Think about what type of graph might be appropriate for which data.

Proportional graphs

Pie charts are a good method of displaying data proportionality. A pie chart is divided into segments with angles proportional to the data (Figure 14). The pie circle represents 100% of the data set.

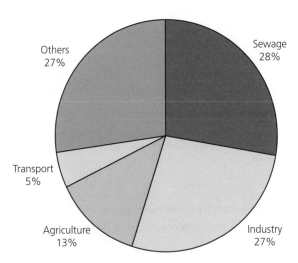

Note: 'Others' are incidents in which the source cannot be accurately identified.

Figure 14 River pollution incidents in England and Wales, 1990

Sometimes the size of each pie chart is used to indicate the size of a population, and this can be compared to a scale.

> **Exam tip**
>
> Always have a calculator, pencil, ruler, colouring pencils and an angle measure with you for every geography exam.

Scattergraphs

Scattergraphs are slightly more complicated because they often involve two different types of observation. Each observation is called a variable. Scattergraphs plot the relationship or correlation between the two variables (see Figure 3, page 13).

Once the graph is drawn the strength of the relationship can be tested using a 'line of best fit'. The graph might then look like one of four possibilities (Figure 15).

(a)
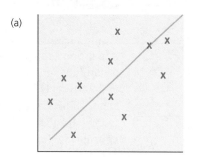

(b)
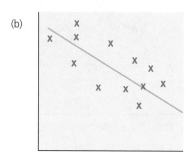

(c)
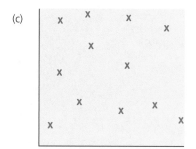

(d)
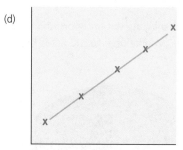

Figure 15 Scattergraphs showing (a) strong positive correlation, (b) strong negative correlation, (c) no correlation, (d) perfect positive correlation

The drawing of a scattergraph is usually the starting point for the Spearman's rank correlation statistic.

Triangular graphs

Triangular graphs are plotted using three axes connected in an equilateral triangle (Figure 16). It is only possible to show three variables and each component must be measured out of 100%.

Exam tip

Scattergraphs can take time to perfect, so make sure that you practise drawing them.

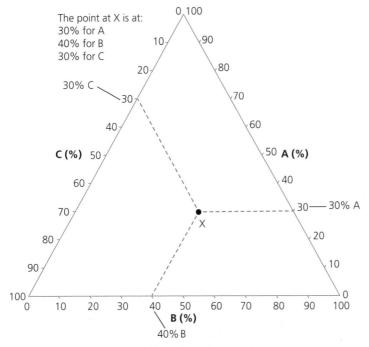

Figure 16 A triangular graph

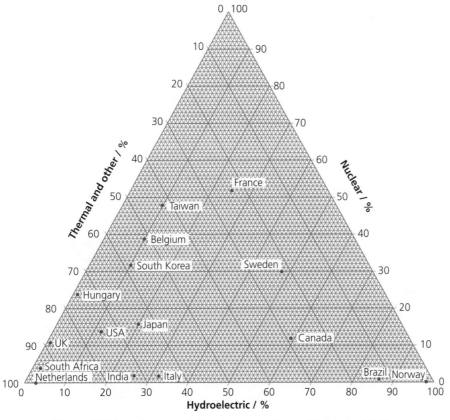

Figure 17 Triangular graph showing percentage of total electricity production by generating source for selected countries (1988)

Sampling methods

Random sampling is when a random number table or random number generator app is used to give the sequence of, for example, people to ask or houses to call in a street when doing a survey.

+ Using a random number table means that results will be totally random and should be unbiased — this should also give a good representation of the whole area/population.

− The technique takes no account of any changes/subsets/differences in an area or population, so a population survey could end up asking only people aged over 65 (this is called 'bunching').

Systematic sampling is when samples are taken using a pre-determined interval. Questionnaires might target every 5th or 10th person who walks past, soil studies might be taken every 5 m or 10 m along a survey line/transect.

+ This is best for studies that are measured over time or over a particular distance. It is suitable for studies like sand dune surveys or measurements taken across a river.

− There is an increased chance of bias as it is the individual researcher who decides on the interval. Fixed intervals might mean that important measures (like the slope angle on a sand dune) might be meaningless, as the upward or downward slopes might not appear in the results.

Stratified sampling is useful when there are clear sub-groups in the dataset. For example, if conducting a sample of 70 questionnaires in a school population of 700 students, you might break the sample down so that equal numbers of questionnaires are given to each year group — 10 to year 8, 10 to year 9, 10 to year 10 etc. You could break this down even further and ask for two questionnaires from each of five form classes and again break things down to ask one boy and one girl from each form class.

+ This can give a broad and detailed snapshot of what a population/group thinks about a particular issue.

− This can be a complicated and difficult method of collecting results, taking a lot of time and organisation.

Point sampling is when individual points are used in the investigation. For example, specific accessible sites might be chosen in a river study.

Line or **transect sampling** is when a line is drawn on a map in an area and all data are collected along this line.

Quadrat sampling is when a piece of equipment called a quadrat is used to measure the amount of vegetation/type of vegetation or amount of ground coverage within an area. The quadrat is usually a square metal frame (the most common is 50 cm × 50 cm), which is placed onto the ground during an ecological study.

Pragmatic sampling is when decisions are taken to visit sites that are safe/accessible or which might demonstrate typical characteristics. Although this approach often allows for a simple fieldwork experience, it introduces a huge amount of bias into the sampling technique.

Exam tip

Make sure that you know the positives and negatives of using each of the different sampling methods. Each one will be more useful for a particular type of investigation.

Methods of statistical analysis

Many students struggle with the statistical analysis section of AS geography but these different statistical techniques are relatively straightforward to use and understand.

Mean, median, mode and range (measures of central tendency)

The **mean**, often known as the average, is found by adding together all the values under investigation and dividing this by the total number of values.

+ It can be an accurate measure as all values are seen as equally important, and it is a relatively simple calculation.

– It can be distorted if there is one extreme value, and sometimes the decimal places can cause confusion. For example, if a total fertility rate is 2.3, how can the average mother have 0.3 of a child?

The **median** is the central value when the values are ranked in a series. If there is an even number, the median is the mid point between the two middle values.

+ It is a good way of finding the 'centre' of the distribution pattern within the data set and is not usually affected by extreme values.

– It is less reliable when there are few values and it cannot really be used for any further mathematical processing.

The **mode** is the most frequently occurring number in the data set.

+ It is quick to calculate and can help to describe the general distribution of the data.

– It has limited value as the modal value could end up being one of the extreme values, which would cause bias in the result set. It does not take into full consideration the full range of values in the data.

The **range** is the difference between the highest value and the lowest value in the set of numbers. It shows the spread of the data.

Spearman's rank correlation

Drawing a scattergraph is often the first step in trying to understand the relationship between two variables or two groups of data in a fieldwork study.

The Spearman's rank correlation coefficient (r_s) is a statistical test that shows the strength of the correlation or relationship between the two groups of data. It shows the strength and type of relationship.

Usually, the Spearman's rank correlation only works when there are at least 15 sets of values. The first step is to think of a hypothesis to test. In this example we investigate whether 'there is a negative correlation between global birth rates and GNI per capita'.

Knowledge check 11

Each student in a class of 19 students noted the age of their eldest surviving grandparent:

90, 89, 80, 76, 65, 54, 85, 75, 81, 76, 73, 72, 66, 74, 69, 70, 64, 62, 77

(a) What are the mean, median and mode for these data?

(b) Which of these measures do you think is the most useful, and why?

(c) Which of these measures do you think is the least useful, and why?

Table 3 Spearman's rank correlation for birth rate against GNI

Country	Births per 1,000 population	Rank	GNI PPP per capita (US$) 2010	Rank	d	d^2
Burkina Faso	43	2	1,250	15	−13	169
Kenya	35	3	1,640	13	−10	100
Zambia	46	1	1,380	14	−13	169
Tunisia	19	7.5	9,060	10	−2.5	6.25
South Africa	21	5	10,360	9	−4	16
Canada	11	14	38,370	1	13	169
Mexico	20	6	14,400	7	−1	1
Jamaica	16	9.5	7,310	11	−1.5	2.25
Argentina	19	7.5	15,570	6	1.5	2.25
Brazil	16	9.5	11,000	8	1.5	2.25
Bangladesh	23	4	1,810	12	−8	64
Iceland	14	11.5	28,270	4	7.5	56.25
Italy	9	15	31,810	3	12	144
UK	13	13	35,840	2	11	121
New Zealand	14	11.5	28,100	5	6.5	42.25
						$\Sigma = 1,064.5$

Source: figures from the 2012 World Population Data Sheet (Population Reference Bureau, 2012)

Step 1 The formula for investigating Spearman's rank is:

$$r_s = 1 - \left(\frac{6\Sigma d^2}{n^3 - n}\right)$$

Step 2 The next stage is to work out the rank order of the values within each variable (from the highest value to the lowest). When there are two values the same, add both rank values together and divide by 2. For example, Jamaica and Brazil both have a birth rate of 16 and they should be ranked 9 and 10 on the table. Add 9 + 10 and divide by 2 = 19/2, so both are ranked as 9.5.

Step 3 Calculate the difference (d) between each of the pairs of rank values (first value minus the second value).

Step 4 Square each of the resultant values (d^2). Note that all minus numbers should disappear at this stage.

Step 5 Add all the d^2 numbers to get the total (Σ) for d^2.

Step 6 Go back to the formula and use the data to complete the calculation. Always remember to show your working so that you can score marks even if you get the wrong final answer.

$$r_s = 1 - \left(\frac{6\Sigma d^2}{n^3 - n}\right)$$

$\Sigma d^2 = \dfrac{\text{total}}{\text{sum of the differences in the values of each matched pair squared}}$

n = the number of different values measured for comparison (ranked pairs)

$$r_s = 1 - \left(\frac{6 \times 1{,}064.5}{15^3 - 15}\right) = 1 - \left(\frac{6{,}387}{3{,}375 - 15}\right) = 1 - \left(\frac{6{,}387}{3{,}360}\right)$$

$$r_s = 1 - 1.9 = -0.9$$

Step 7 Check that the final value lies between −1 and +1. If it does not then you have done something wrong in your calculation.

Step 8 Comment on the strength of the relationship — any result between 0 and −1 is seen as having a negative correlation. The closer the result is to −1, the stronger the correlation. Any result between 0 and +1 is seen as having a positive correlation. The closer the result is towards +1, the stronger the correlation. In this case, the relationship has a very strong negative correlation.

Step 9 Comment on the statistical significance. In the exam you will be given both a graph and a table to help you to work out the significance of your result (Figure 18).

When investigating correlation there is always the possibility of bias in collating the data or of a result occurring *by chance*. If there is a possibility that chance occurred within more than 5% of the data, then this is considered unacceptable and we should not accept the results. This is sometimes called the 5% rejection level.

Using the graph, plot the degrees of freedom on the horizontal axis by counting the number of pairs (*n*) minus 2. Then plot the Spearman's rank result on the vertical axis.

- If the value lands above the 99.9% significance level line, this means that the relationship is strongly significant and there is a probability of less than 1 in 1000 that the relationship occurred by chance.
- If the value lands between the 99% and 99.9% significance lines, this means that there is a probability of less than 1 in 100 that the relationship occurred by chance. We can still accept the relationship and result.
- If the value lands between the 95% and 99% significance lines, this means that there is a probability of up to 5 in 100 that the relationship occurred by chance.
- If the value lands on or below the 95% significance line, this means that there is a probability of at least 5 in 100 that the relationship occurred by chance and we find the result to be non-significant. We should reject any hypothesis, as the element of chance or bias interfering with the results is too much.

Step 10 Provide a geographical explanation — a final question part might ask you to 'give geographical reasons that could be suggested to explain this statistical result'. To answer this you need to go back to the original hypothesis and the original variables that are being tested and use the Spearman rank result, strength and level of significance to show how strong the relationship is.

In this case, the result is a strong negative correlation (and strongly significant at the 99.9% level), which means that as one variable increases the other decreases. In this case, as the amount of GNI per capita increases the birth rate is expected to decrease.

Your explanation might then go on to explore reasons why people who live in a country where GNI is very high might then have enough money for contraception, or might be focused on careers instead of family size, with the result that birth rates are much lower than in countries with a lower GNI.

Exam tip

Spearman's rank is relatively straightforward when you know how. At least one of the statistical methods is likely to be examined at length in Unit 3. Know how to use them and, more importantly, know how to talk about the significance and how to generate a geographical reason to explain a correlation.

Knowledge check 12

(a) Describe the statistical significance of the results.

(b) Give geographical reasons that could be suggested to explain this statistical result.

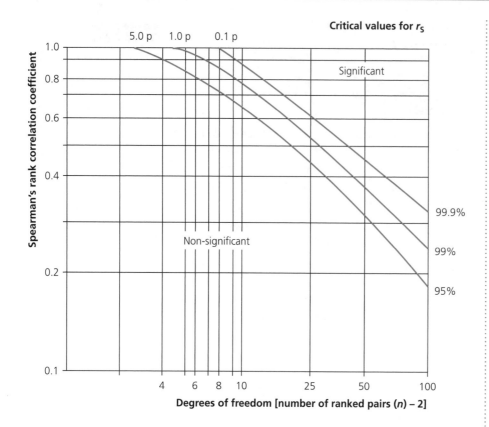

Figure 18 Spearman's rank correlation significance graph and table

Degrees of freedom	Significance level	
	0.05 (5%)	0.01 (1%)
8	0.72	0.84
9	0.68	0.80
10	0.64	0.77
11	0.60	0.74
12	0.57	0.71
13	0.54	0.69
14	0.52	0.67
15	0.50	0.65
20	0.47	0.59

Nearest neighbour analysis

Many investigations involve the arrangement of spatial data, sometimes called 'measuring dispersions'. On maps, settlements often appear as dots. Nearest neighbour analysis (R_n) is a statistical measure that allows us to look at a distribution (usually a dot map) and to determine a pattern.

Nearest neighbour analysis produces a figure to show the extent to which the pattern on the map (Figure 19) is found to be clustered (nucleated), random or regular (uniform).

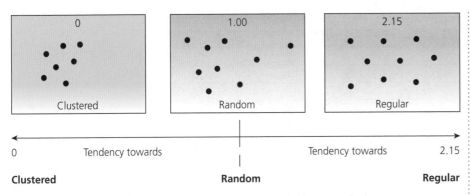

Figure 19 Patterns for nearest neighbour analysis

The basis of the statistic is to measure the distance between each of the dot points on a particular map.

Step 1 The formula for investigating nearest neighbour is:

$$R_n = 2d\left(\sqrt{\frac{n}{a}}\right)$$

where d is the mean distance between the nearest neighbours, n is the number of points and a is the area under study.

Step 2 The first part of the formula is found out by using the map to work out which of the points is the nearest to each and to measure the distance between them. This should be measured in km. (See, for example, Figure 20 and Table 4.)

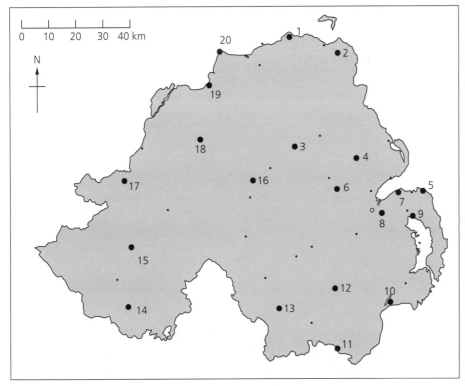

Figure 20 Map of Met Office automatic synoptic and climate stations, 2012

Table 4 Nearest neighbour measurements for Met Office stations in NI

Met Office station	Number	Nearest neighbour	Distance (km)
Giant's Causeway	1	2	25
Ballypatrick Forest	2	1	25
Portglenone	3	16	20
Killylane	4	6	12
Orlock Head	5	9	10
Aldergrove	6	4	15
Helens Bay	7	9	9
Stormont Castle	8	7	8
Ballywatticock	9	5	7
Murlough	10	12	22
Killowen	11	12	20
Katesbridge	12	10	22
Glenanne	13	12	25
Derrylin	14	15	25
St Angelo	15	14	25
Lough Fea	16	3	20
Castlederg	17	15	27
Banagher	18	19	22
Ballykelly	19	20	12
Magilligan	20	19	12

Map area = 175 km × 150 km = 26,250 km^2

Step 3 Calculate the mean distance (d) between the nearest neighbours by adding all the distances and then dividing by the number of points/places. In this case the total is 363 km, giving a mean of 363/20 = 18.15 km.

Step 4 Go back to the formula and use the data to complete the calculation. Always remember to show your working out so that you can get marks even if you get the final answer wrong.

$$R_n = 2d\left(\sqrt{\frac{n}{a}}\right)$$

$$R_n = 2 \times 18.15 \times \sqrt{\frac{20}{26,250}}$$

$$R_n = 36.30\sqrt{0.0007}$$

$$R_n = 36.30 \times 0.026$$

$$R_n = 0.94$$

Step 5 Check that the final value lies between 0 and 2.15 — if it does not then you have done something wrong in your calculation.

Step 6 Comment on the pattern. On the exam paper you are given a graph to help you comment on the R_n value (Figure 21). A result towards 2.15 is regular while a result towards 0 is clustered. A result that is around 1 is random, but might have a tendency towards regularity or clustering.

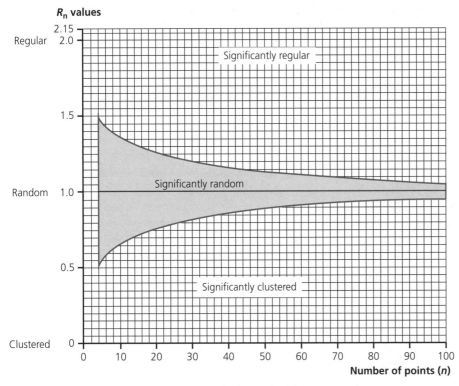

Figure 21 Nearest neighbour significance graph

Step 7 Comment on the statistical significance — the graph also helps you to work out the significance of your result. Plot the number of points along the horizontal axis and then read the R_n result up the vertical. This tells you if the result is significantly random, clustered or regular.

In this example, the distribution pattern of the weather stations is nearly perfectly, significantly random. There is a tiny tendency towards clustering but there is no real regularity at all. In many cases, a result that is random is considered to be at the 95% significance level, which means that there is a probability of 5 in 100 that the relationship occurred by chance, and so we reject the result and any hypothesis.

Step 8 Provide a geographical explanation — a final question part might ask you to 'give geographical reasons that could be suggested to explain this statistical result'. To answer this you need to explain why the location might be clustered, random or regular. In this example, the location of weather stations is random, though there is the slightest of tendencies towards clustering, which might be explained by the clustering of a few stations in and around Belfast.

Limitations of using nearest neighbour analysis

Nearest neighbour analysis has a number of limitations that might create bias in the result:

- Mathematicians state that a minimum of 30 points is needed to ensure reliability and validity of conclusions.
- The size and scale of the area under investigation are extremely important. A larger area will lower the R_n value and will show more clustering. A smaller area will increase the R_n value and the pattern will be more regular.

Knowledge check 13

(a) What evidence might there be to show that this result of 0.94, is unreliable?

(b) What might be the impact on the R_n value if the area were changed to just look at the County Antrim area?

Exam tip

Nearest neighbour analysis is often used for settlement studies, but be aware that it can have different uses. Make sure that you are aware of the different limitations of the technique.

■ It is often used to measure settlements. Which settlements should be included? Do individual isolated buildings or hamlets count? For larger towns and cities there can be confusion about where the central area is, i.e. where the distance should be taken from.

Summary

- Photos and maps can be used for students to draw an annotated sketch.
- Dot maps can be used to illustrate distribution patterns in an area.
- Flow line maps can show the strength of a movement in a direction.
- Choropleth maps are visually appealing but can be difficult to draw.
- Isoline maps use lines to join areas with the same value.
- Bar graphs, line graphs, scattergraphs, proportional graphs and triangular graphs can all be used to present geographical data.

- Geographers must choose an appropriate sampling technique to collect their data (random/stratified/systematic/pragmatic).
- Mean, median, mode and range are simple statistical measures.
- Spearman's rank correlation measures the relationship between two sets of variables.
- Nearest neighbour analysis allows measurement of the pattern of dispersion in an area, ranging from clustered to regular spacing.

A2 Unit 3

■ Decision-making in geography

Key attributes of the decision-making process

The compulsory decision-making examination paper in A-level geography is often feared by teachers and students alike. However, careful planning, practice and preparation can ensure that students can actually enjoy this experience!

The study of geography allows students to consider the different connections and inputs that are involved in planning and making decisions. This unit provides students with a real world scenario in the form of a case study and asks them to identify and analyse the resources, examine conflicting views and make decisions and justify any recommendations.

The decision-making paper will provide you with information about a real world scenario, presented as a resource booklet. This will include information and resources from a wide range of sources including: maps (at a variety of scales), photographs, annotated diagrams, tables, statistical data, reports and infographics, general information, arguments in favour of the proposal, arguments against the proposal and quotations that might be related to the development.

Students will have 30 minutes to read through, analyse and process the information so that they can respond to the decision-making task. There are a further 60 minutes remaining to respond to the demands of the question and to come up with a final decision for the proposal.

Exam tip

The format of the decision-making paper has not changed much over the years — you can still practise using the papers from the old version of the CCEA exams. These are usually found with the A22 Physical exam papers up to 2017.

The issues that make up the decision-making exercise usually come from anywhere in the world but usually do not focus on Northern Ireland.

Some of the recent decision-making exercises have included:

- 2015   A potash mine in the North York Moors National Park
- 2014   A port development plan for Falmouth
- 2013   The development of an airport in St Helena
- 2012   A hillside windfarm development in Scotland

Using the resources

There is not a lot of time to spare when trying to read through all of the resources in the task.

Understanding the issue

Usually, the decision-making paper will clearly identify the main context of what the issue is through a 'Background to the Report'. This will contain a basic statement of the problem and will usually also identify the role that you, as the writer of the decision-making report, will have to take for the issue.

Adopting a role

One of the key aspects of the written report is that students need to adopt and maintain a particular role. This role usually is for someone who would be in a position to either recommend that a particular project continue or be stopped. The idea here is for the student to consider the evidence and then to come out with an opinion as to what they think needs to happen next. It is important that candidates try to maintain this role through their written reporting of the evidence.

Sorting through the evidence

There is a lot of evidence and material for each candidate to wade through in this exam. It is important to process this information and to be able to sift the information and decide what is a useful element for your argument.

You should first read through the 'Background to the Report' carefully so that you understand the main focus of the issue and the role you have to take fully. Then, you should take a look at the individual information that suggests the way that your report should be structured. Make sure that you know the main areas that your report is going to need to focus on.

Some students find it useful to take brief notes on separate blank pages in favour of or against the proposals. Another idea to help you sort the resource information is to use a selection of highlighter pens to code the information as being relevant to each section of the report — e.g. yellow for introduction, green for environment, pink for the economy and blue for social aspects.

Think: What does the evidence presented tell you about the issue? What does it show? What data is given in support of/against the issue? Does the evidence lead you to think that the proposal is worth supporting or going against? What type of evidence are you looking at? Is the evidence based on reliable facts, values, experiences or is this just based on one person's opinion? How far can you trust the validity of the evidence?

> **Exam tip**
>
> You need to make sure that you have clearly shown the role that you have to adopt in at least two places through the report — make this as obvious as possible.

Making a decision

It is never easy to make a decision about whether you agree or disagree with the proposals presented. Some students make the mistake of trying to sit on the fence — but it is important that as you adopt a role, you also come out in favour of or against the proposal. Once you have done this you need to make sure that you clearly give both sides of the argument and also clearly justify why you think your final decision is valid. Do not doubt yourself.

Tips and techniques

Writing up the report

You have just less than 1 hour to use this information to formulate your argument so that you can score the maximum number of marks for this section. You need to make sure that you plan your answer carefully using the resources provided. Spend a few minutes planning your answers for each of the sections before you begin and try to identify which resources you are going to use to help prove your case in each section.

Managing your time

In this exam, timing is everything. You need to make sure that you spend the right amount of time on each section of the written report and leave yourself enough time to complete all the aspects of the report.

Often the question will ask you to make a decision on the basis of 'the greater overall benefits'. To complete this you need to give credit to any opposing arguments but you then need to explain why you have set these aside to go with an alternative decision.

The exam board notes that in order to avoid spending too much time on any particular section, candidates should:

1 answer quickly, succinctly and keep content relevant

2 focus on the elements of the resource relevant to that section

3 read the question carefully to avoid straying into material that you will need for a later section

Reading the information	30 minutes
Drawing a graph	8 marks (8 minutes)
Writing the report	52 marks (52 minutes) Time (in minutes) should reflect the number of marks for each section
	For example: Introduction (10 marks = 10 mins) Impact on: (1) the economy (10 marks = 10 mins) (2) environment (10 marks = 10 mins) (3) social aspects (8 marks = 8 mins) Final decision (10 marks = 10 mins) (Format and role should be incorporated into the report = 4 marks)

Exam tip

Keeping an eye on your timing is extremely important through the whole exam paper. Have a watch with you on your table and make sure that you have a clear timetable of when you should be starting each section.

Practise, practise, practise

Some students think that there is very little work to be done to prepare for this paper. You do need to carefully consider the advanced statistical analysis (for A2) below which details some of the possible techniques that could be used on a paper like this. In addition, the most important thing is to practise these questions — using the timings suggested and under examination conditions where possible, so that you can get a feel for how to manage your time during this exam.

Advanced statistical analysis (for A2 only)

In order to prepare for the A2 examinations it is a good idea to start by revising the main statistical techniques from the AS exams. The exam specification also notes two further methods of statistical analysis that are required for A2 level only. These are chi-squared and location quotient. Although both techniques might be used on either the A2 1 Physical exam paper or the A2 2 Human exam paper, the highest potential use of these techniques will be on the A2 2 Decision-making paper.

Chi-squared test

In AS geography you will use the Spearman's rank correlation to measure the strength of relationship between two variables. Chi-squared, on the other hand, looks at the differences between the variables/groups or areas. It can be used to assess the difference between what has been observed and what might have been expected (if everything was equal).

It is best to start with a null hypothesis. In the case of the example below the hypothesis is:

'There is no significant relationship between the number of farms and the underlying rock type.'

Area	Rock type	Number of farms
1	Marl	2
2	Chalk	10
3	Sandstone	8
4	Clay	2
5	Limestone	4

Step 1 The formula for investigating chi-squared is:

$$\chi^2 = \Sigma \frac{(O - E)^2}{E}$$

where $\Sigma =$ sum of

$O =$ the observed figure

$E =$ the expected figure

Step 2 In most other geographical statistical techniques, you use the formula once to find the final answer. With chi-squared you have to apply the formula to each area or group of statistics that you are looking at. Usually this means that you take each line of the formula individually and complete a table similar to that below.

	Area	1 Marl	2 Chalk	3 Sandstone	4 Clay	5 Limestone	Total
a	O (observed)	2	10	8	2	4	26
b	E (expected)	5.2	5.2	5.2	5.2	5.2	26
c	(O – E)	–3.2	4.8	2.8	–3.2	–1.2	
d	(O – E)^2	10.24	23.04	7.84	10.24	1.44	
e	$\dfrac{(O – E)^2}{E}$	1.96	4.43	1.50	1.96	0.27	
f	$\sum$ (sum) of	1.96 +	4.43 +	1.50 +	1.96 +	0.27 =	10.12

a   Insert each of the observed figures from the table.

b   If all things were equal across the five areas, how many farms would we *expect* to see in each of the five areas (26/5 = 5.2).

NB: sometimes a chi-squared test might indicate different-sized areas and results can be worked out proportionally.

c   Observed value minus the expected value.

d   The square of the observed minus the expected value (above).

e   The answer above is then divided by the expected value.

f   Finally, the total value above for each of the areas is added together.

This generates our χ^2/chi-squared value.

Step 3 The final result can now be used to find out the significance level. You can use either a graph or a table to help you to work out the significance of your result (Figure 22).

First, work out the number of degrees of freedom – this is done by counting the number of observations minus one. In this case (5 – 1) = 4.

This shows that the result lies between the 1% and 5% level (10.12 is higher than the 9.49 at the 0.05 significance level). In other words, we might expect such a distribution to occur by chance in fewer than five times in every hundred.

Any discovery that is found below the 5% level means that we can reject the null hypothesis and this shows that there is some relationship between the number of farms and the rock type. Any answer found beyond the 5% mark means that we can accept the hypothesis and we can confirm that any relationship does not exist.

Exam tip

Chi-squared is not an easy statistic to use — so make sure that you practise it carefully. In particular, make sure you know how to describe the meaning of the final result.

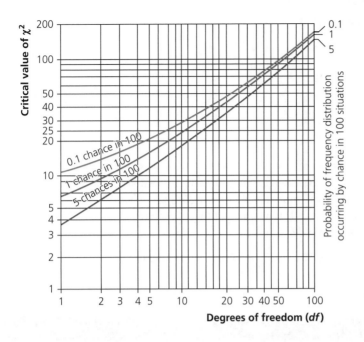

Figure 22 Chi-squared significance graph and table

Degrees of freedom	Significance level 0.05 (95%)	Significance level 0.01 (99%)
1	3.84	6.64
2	5.99	9.21
3	7.82	11.34
4	9.49	13.28
5	11.07	15.09
6	12.59	16.81
7	14.07	18.48
8	15.51	20.09

Location quotient

The location quotient is a measure of the amount to which a geographical activity is concentrated in a particular area. For example it can help to work out the concentration of an industry in an area.

Step 1 The formula for the location quotient is:

$$LQ = \frac{r/n}{R/N}$$

where r = the number of people employed in industry A in area X

n = the number of people employed in area X

R = the number of people employed nationally in industry A

N = the total number of people employed nationally

For example, if there are 7,058 people employed in agriculture in Armagh, Banbridge and Craigavon District Council area out of 90,527 workers in the area, compared with a national total of 48,000 people employed in agriculture across Northern Ireland and a total labour force of 834,000 (figures for 2015).

Exam tip

Location quotient is not the most difficult statistical measure to process but it can be hard to describe so make sure that you practise a few different answers and work out what they actually mean. How could you compare one LQ answer with another and what might that mean, for example how would a LQ score of 0.7 compare to one of 1.5?

Step 2 The location quotient works out as follows:

$$LQ = \frac{r/n}{R/N}$$

$$= \frac{7,058/90,527}{48,000/834,000}$$

$$= \frac{0.078}{0.058} = 1.34$$

If the answer is equal to 1, this means that the number of people who work in that industry is exactly the number that would be expected for that industry, for that area. If the number is above 1, it means that there are more people working in that industry than the national average. If the number is below 1, it means that there are fewer people working in that industry than the national average.

In this case, 1.34 shows that there is a higher concentration of people working in the agriculture industry in Armagh, Banbridge and Craigavon District Council compared to the national average.

Summary

- The decision-making paper in geography requires careful planning and practice.
- Candidates have 30 minutes to read, analyse and process the information and 60 minutes to write their answer.
- Your strategy for sorting through the information will often dictate how successful you are going to be at completing this paper effectively.
- You need to adopt the stated roles and make a final decision about whether the proposal should go ahead or not.
- Careful time management is a fundamental consideration within this task.
- The chi-squared test is another statistical technique that is required for A2 geography students. It helps to measure the difference between what we might expect to see and what we actually observe.
- Location quotient helps us to measure the amount to which a geographical activity is concentrated in a particular area.

Questions & Answers

■ Examination skills

AS Unit 3

The AS Unit 3 geography paper includes the following two questions:

	Compulsory?	Marks (out of 60)	Exam timing (out of 60 minutes)
Section A			
Q1 Fieldwork skills You must take a summary report and a table of data into the exam with you. Questions will be asked in relation to how you continue to process the information from this table and on other fieldwork experiences.	Yes	30	30
Section B			
Q2 Geographical techniques You will respond to qualitative and quantitative data taken from a variety of secondary sources.	Yes	30	30

As with all A-level exams there is little room for error if you want to get the best grade. Gaining a grade A is not easy in AS geography so you need to ensure that every mark counts.

The examination paper for AS Unit 3 is 1 hour long. There are 60 marks available, which means that you get 1 mark per minute to work your way through the paper. You need to make sure that you manage your time carefully. If you find that you have time left over in this exam, the chances are that you have done something wrong.

Exam technique

Students often find it difficult to break an exam question down into its component parts. On CCEA exam papers, the questions are often long and difficult to understand, so you need to work out what the question is asking before you move forward.

A2 Unit 3

The A2 Unit 3 geography paper is made up of a compulsory decision-making exercise which is accessed through the provision of case study information. The unit is structured as follows:

	Compulsory?	Marks (out of 60)	Exam timing (out of 90 minutes)
Decision-making report (example)			
Reading through the information	Yes	0 marks	30
Format of the report	Yes	2 marks	2
Adopt the stated role		2 marks	2
Graph		8 marks	8
Introduction to the report		10 marks	10
Impact on: (1) the economy		10 marks	10
(2) environment		10 marks	10
(3) social aspects		8 marks	8
Final decision		10 marks	10

As with all A-level exams there is little room for error if you want to get the best grade. Gaining a grade A* is not easy in A2 geography so you need to ensure that every mark counts.

The examination papers for A2 Unit 3 last for 1 hour 30 minutes. You should use about 30 minutes of the exam time to read through the resources given. There are 60 marks available, which means that you get 1 mark per minute to start to write your report using the information provided for the remaining 60 minutes. The main reason why so many students struggle with this paper is that they fail to manage their time appropriately and as a consequence they do not have enough time left to answer the essays at the end in sufficient detail. If you find that you have time left over in this exam, the chances are that you have done something wrong.

Information for both papers

Command words

To break down the question properly, get into the habit of reading the question at least *three* times. When you do this it is sometimes a good idea to put a circle round any command or key words that are being used in the question.

A common mistake is failing to understand the task being set by question. There is a huge difference between an answer asking for a discussion and one asking for an evaluation.

The main command words used in the exam are as follows:
- **Compare** — what are the main differences and similarities?
- **Contrast** — what are the main differences?
- **Define** — state the meaning (definition) of the term.
- **Describe** — use details to show the shape/pattern of a resource. What does it look like? What are the highs, lows and averages?
- **Discuss** — describe and explain. Argue a particular point and perhaps put both sides of this argument (agree and disagree).
- **Explain** — give reasons why a pattern/feature exists, using geographical knowledge.
- **Evaluate** — look at the positive and negative points of a particular strategy or theory.
- **Identify** — choose or select.

■ About this section

Two sets of practice exam questions with exemplar answers are provided for AS Unit 3 Question 1 and Question 2, and one for A2 Unit 3. These will help you to understand how to construct your answers in order to achieve the highest possible marks.

Commentary

Some questions are followed by brief guidance on how to approach the question (shown by the icon ⓔ). Student responses are followed by comments, preceded by the icon ⓔ, indicating where credit is due. In the weaker answers, they also point out areas for improvement, specific problems, and common errors such as lack of clarity, weak or non-existent development, irrelevance, misinterpretation of the question and mistaken meanings of terms.

AS Unit 3

■ Question 1 Fieldwork skills

Two questions have been supplied in this section to demonstrate the different types of questions that you might encounter in the exam; each question is worth 30 marks.

Question 1A

(a) Study the points below, which outline some important considerations made by a student when preparing for a geography fieldwork trip:

- Transport to the site
- Accessibility of the site
- Safety equipment
- Suitable clothing for fieldwork
- Communication devices

Select one of the planning considerations above and discuss its importance and role within your fieldwork.

(3 marks)

ⓔ 3 marks are awarded for an answer that deals with both the importance and the role of the selected factor and makes a convincing case, with appropriate reference to the individual fieldwork. 1–2 marks are given for a more simple response, which might fail to address either the importance or the role, with unconvincing reference to the individual fieldwork.

Student answer

(a) When completing a fieldwork project in the sand dunes we had to think about specific safety equipment that would keep us safe on the site visit. We were worried about bad weather, so our teacher told us to make sure that we had a waterproof coat, waterproof trousers and a hat and gloves to keep us warm (it was a very cold, wet and windy day — so this advice was really important!). Also, we had to wear walking boots as the terrain was rough and this would give us more support. We brought waders and a safety throw rope as we were going to have to collect water from the sea to help with our experiment.

ⓔ 3/3 marks awarded. This makes reference to three different aspects of safety equipment that were used for this particular fieldwork. It discusses both the role of the equipment and the importance of this within the context of their visit to the sand dune.

(b) Describe and explain **one** sampling method that was used within your fieldwork.   (6 marks)

ⓔ You can select from a range of different sampling methods, but most relate to one of either random, systematic, pragmatic or stratified sampling. 3 marks are for a general description of the sampling method and the other 3 marks are for an explanation of how the sampling method works in the context of this particular fieldwork.

(b) The aim of our fieldwork was to visit a total of 15 sites across the sand dune ecosystem at Magilligan Point. We were trying to see how the infiltration rate (how quickly 200 ml of water soaked into the soil) changes as you move away from the sea.

Our sampling method was to use systematic sampling to take measurements at sites along a line stretching back from the sea through the sand dune system. We took measurements at 15-m intervals along this line. We did this as other sampling techniques would not have been right for this coursework. We wanted to see how the infiltration changed through the sand dune and not just at random places. Pragmatic sampling might have worked too as the problem with systematic sampling was that sometimes the sites were not showing the changes that actually happened in the sand dune.

ⓔ 5/6 marks awarded. This is a good description of how the sampling technique was used within the context of this fieldwork. There is good depth about how systematic sampling was used to look at the changes across the sand dune system. The answer then goes on to explain why this sampling method was appropriate and even mentions how another sampling method might have been used. However, this explanation could have been more focused.

(c) (i) Choose **one** of the following statistical techniques that could be used to analyse some of your fieldwork data. Your chosen technique must fit in with the aim of your fieldwork.
  ■ Spearman's rank correlation
  ■ Nearest neighbour analysis
  ■ Mean, median, mode and range
  In the space provided, complete your chosen statistical analysis and show all your calculations clearly. If relevant, comment on the level of statistical significance. (Significance graphs and formulae are provided — see pages 11–12 and 32–33.)   (7 marks)

ⓔ The statistical technique depends on the chosen fieldwork, but it needs to be relevant to the aim/hypothesis of the investigation.

(c) (i) Statistical technique selected: Spearman's rank correlation

Site (distance from the sea in m)	Rank	Time for infiltration/s	Rank	d	d^2
0	15	10.1	15	0	0
15	14	32.3	14	0	0
30	13	62.5	12	1	1
45	12	95.5	11	1	1
60	11	120	10	1	1
75	10	133	8	2	4
90	9	62	13	−4	16
105	8	156	7	1	1
120	7	132	9	−2	4
135	6	171.4	6	0	0
150	5	235	5	0	0
165	4	245	4	0	0
180	3	266	1	2	4
195	2	256	3	−1	1
210	1	262.5	2	−1	1
					$\sum d^2 = 34$

$$r_s = 1 - \left(\frac{6\sum d^2}{n^3 - n}\right)$$

$$r_s = 1 - \left(\frac{6 \times 34}{15^3 - 15}\right)$$

$$r_s = 1 - \left(\frac{204}{3{,}375 - 15}\right)$$

$$r_s = 1 - \left(\frac{204}{3{,}360}\right)$$

$$r_s = 1 - 0.06$$

$$r_s = 0.94$$

This shows a strong positive correlation between the two variables. The result is significant at the 99.9% level. This is a very significant result.

🅮 **7/7 marks awarded.** This student has opted to complete a Spearman rank correlation using the two data variables that are indicated in the aim of the fieldwork. The table is adapted for Spearman rank data. Full working out of the formula is shown, with a final Spearman rank result. The answer then comments on the relationship and on the significance of the result. There are no errors or inaccuracies here and the organisation of the task is logical, answering all aspects of the question.

(c) (ii) Explain your statistical outcome with reference to relevant geographical theory or concepts.

(6 marks)

ℯ The answer must be based on *geographical* reasoning in relation to the calculated statistical result. The reasoning will depend on the fieldwork undertaken. There should be evidence of geographical theory or concepts to support the interpretation.

Level 3 (5–6 marks): Sound, relevant, geographical concepts are presented and discussed in an effective manner, using specialist terminology in written communication.

Level 2 (3–4 marks): An accurate discussion, but the answer may lack depth. There may be less evidence of geographical theory.

Level 1 (1–2 marks): Explanation is basic and lacks depth.

(c) (ii) Having completed the Spearman's rank correlation, I found that $r_s = 0.94$. This means that there is a strong positive correlation between the two variables. The result is significant to the 99.9% level, which means that the relationship did not occur by chance.

There is a geographical reason that connects the two variables. As you move away from the sea and move through the sand dune the soil structure in the sand dune begins to get more complex. This is because the vegetation on the sand dune gets more complicated — at the front of the sand dune the only plant that can survive is marram grass and this starts to hold the sand/soil together, but as you move further back into the sand dune ryegrass and moss start to develop and these help to add nutrients and humus into the soil and the soil can start to retain more moisture, making the infiltration rates slow down the further back that you go.

ℯ **5/6 marks awarded.** This is a good answer, which describes the relevance of the statistical technique in testing the relationship within the stated aim/ hypothesis. There is also good reference to geographical concepts and theory to back up the technique used. Specialist terms are used and specific knowledge in relation to seral succession is shown.

(d) Describe and evaluate **one** way in which you could extend your fieldwork study to explore your aim further and improve your geographical knowledge.

(4 marks)

ℯ Answers will vary depending on the fieldwork chosen. 2 marks are awarded for a valid description of a valid possible extension to the fieldwork. The other 2 marks are for an evaluation of how this extension might improve the fieldwork experience.

(d) The main aim of our fieldwork was to look at the relationship between distance from the sea and the infiltration rate. We realised when we had completed the 15 sites that we had really only gone about halfway through the sand dune. We had not made our way through all of the different changes in the sand dune and had not reached the climax vegetation levels. Therefore, to improve our fieldwork we should have continued until we had reached the back of the sand dune — though this might have meant another 20 site measurements, which would have taken a long time. However, we would have had a lot more data with which to make sure that our hypothesis was accurate.

ⓔ **4/4 marks awarded.** This makes clear reference to the aims stated in the report and then describes how the fieldwork could be improved further by visiting additional sites. The answer is developed further with some evaluation of how this might be positive (improving accuracy) but also negative (a lot more sites needed = more time).

(e) Describe **one** way in which the place where you carried out your fieldwork proved to be suitable for carrying out the stated aim in your fieldwork report. (4 marks)

ⓔ You are expected to reflect on the location of your fieldwork (as noted in the report) and should attempt to justify the suitability of the place in the light of the stated aim. You should note the aim and link this to a discussion about how suitable the place was. 3 or 4 marks are awarded for a valid, well-argued answer that demonstrates good knowledge of the chosen fieldwork site and links this with the stated aim/hypothesis. A more limited answer, for 1 or 2 marks, might be down to failure to discuss the suitability of the site or link it to the aim/hypothesis.

(e) In our study we were aiming to see if there was a link between the infiltration rate and the distance that you move away from the sea into a sand dune. In class we had learnt about the different changes that a sand dune goes through as it develops and so we looked for a sand dune system that would allow us to see some of these changes. Magilligan Point was a really good dune to study as we were able to see pictures of what the dune used to look like 200 years ago and we could see how this was different from today. The sand dune system was accessible and was a safe environment for us to record our results.

ⓔ **3/4 marks awarded.** There is some good depth to this answer. It refers well to the aim of the study and explains why this particular location might have been suitable, but to get the final mark it should have made the link between the aim and the location more detailed. For instance, why was this a better location for taking infiltration measurements than another place? Conversely the answer could have argued why using a different location would have given better or more accurate results.

Question 1B

(a) With reference to sampling and risk assessment, discuss how these might be considered by a geographer when planning a field study. (6 marks)

e 1 mark is awarded for a general understanding of the task, 1 mark for a clear and specific link to fieldwork, and 1 mark for an explanation of how the task was completed. Each task is worth 3 marks.

Student answer

(a) Sampling: in our fieldwork study at Magilligan Point, Co Derry, we used a systematic sampling technique. To do this we identified 15 different sites that we would sample. These sites went from before the foredune through to the grey dune. Each site was 15 metres further from the sea than the last site.

Risk assessment: for our fieldwork we identified the key risks at Magilligan Point, for example the lack of shelter and strong winds from the sea. The local weather forecast was then checked before our research study so that we could bring suitable equipment and clothing to ensure safety — e.g. waterproof coats and gloves. Risk assessment was also carried out by identifying key hazards on the ground and making sure that people carried the equipment safely.

---

e **6/6 marks awarded.** Both tasks are dealt with in good depth. A general understanding of both tasks is shown and a clear link is developed specific to the fieldwork outlined in the fieldwork report. There is some discussion about how the sampling and the risk assessment were carried out in this particular case. This answer is worth the maximum 3 marks for each task.

(b) Describe **one** of the primary data collection techniques that were used in your fieldwork to produce the results in your submitted table. (3 marks)

e You must select and write about a primary source of data collection that is referenced in your submitted table. 3 marks are awarded for a detailed description of the data collection technique with explicit and detailed reference to the equipment used (if relevant). 1 or 2 marks are awarded if the description of the techniques lacks depth and reference to fieldwork.

(b) To measure the time taken for 200 ml of water to infiltrate the soil we needed an infiltrometer, 200 ml of water, a measuring cylinder and a stopwatch. The vegetation was cleared so that we were measuring the time taken for the water to infiltrate the soil and not the vegetation. It was then firmly placed in the soil so that water did not escape through the sides. 200 ml of water was then poured in and the time taken for it to infiltrate was recorded. The data was collected accurately as we did it 3 times and took an average.

ⓔ **3/3 marks awarded.** This is a good description of the technique used in the fieldwork. It makes explicit reference to the equipment used to generate the fieldwork table and shows how the equipment was used. This answer scores full marks.

(c) (i) Using some, or all, of the information from your fieldwork table and report, draw a graph relevant to the aim of your fieldwork. (7 marks)

ⓔ Your graph must be accurate and relevant to the aim of the fieldwork, and you must use data from your fieldwork table. 1 mark is available for the title: you must state clearly each of the variables presented. 2 marks are for the conventions: axes must be labelled (variables and units of measurement), a key must be included and scaling (if appropriate). 3 marks are available for accuracy: precision of values, and 1 mark is available for the method: selection of appropriate graph for the task. (Graph paper will be provided in the exam.)

(c) (i)

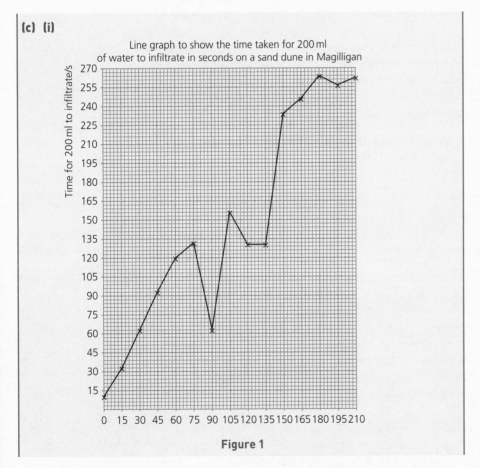

Figure 1

ⓔ **5/7 marks awarded.** The title is long but it generally describes the content of the graph. Conventions have been followed but one of the axes is not labelled fully so only 1 mark can be scored. Some of the values are not as precise as they could have been and this means only 2 marks can be awarded for accuracy. The method selection (a line graph) is acceptable. Take care to draw graphs quickly and carefully.

(c) (ii) Describe and explain **one** geographical factor that might have influenced the results displayed on your graph. (4 marks)

ⓔ Again, the answer will depend on the particular fieldwork studied. The factor selected must be geographical and can relate to either human or physical factors. 4 marks are available for a thorough geographical discussion (interpretation of the graph) that clearly refers to and explains the results generated in the graph. 1 or 2 marks will be awarded for a less-well-argued discussion.

> **(c) (ii)** The amount of vegetation. As you move back through the psammosere the amount of vegetation became increasingly thick and dense. It became very difficult to remove the vegetation to measure the infiltration rate. At some of the sites it may have been impossible to fully clear the site of vegetation and this could have increased the time taken to infiltrate. The more complex the vegetation through the dune, the more humus and leaf litter will also make the soil more complex.

ⓔ **3/4 marks awarded.** This is a decent attempt to describe and explain the role that vegetation might have played in the development of the ecosystem and how the vegetation has made an impact on the infiltration rate. The answer goes into some depth but could make more reference to how vegetation had affected the actual infiltration rates at each site.

(d) Select two of the factors shown below and explain how they might have influenced your results and final conclusions.
- Time of year
- Fieldwork equipment
- Weather conditions
- Time of day
- Group organisation
- Human influence (6 marks)

ⓔ You need to evaluate the influence that each of two factors might have played in generating the results and final conclusions to the fieldwork. 3 marks are available for one factor if the argument is coherent and shows a good understanding of how the factor influences results and conclusions. The answer should clearly reference the fieldwork. 1 or 2 marks will be awarded for a less commanding answer that might not explain the influence of both results and conclusions, or might not reference the fieldwork.

> **(d)** Fieldwork equipment: in our fieldwork study we were measuring the infiltration rate across the sand dune. The infiltration ring that we were using was very basic and sometimes it was used in the wrong way. Our teacher advised us to clear the vegetation and then try to sink the ring into the sand/soil as much as possible but this was not always easy and sometimes water leaked out of the side and this might have affected the

results and the conclusions. It would have been better to use a steel, double-ringed infiltrometer as this might have given more accurate results. Human influence: sand dunes show lots of human influence — there are paths, signs, litter and even an old castle in the dunes. There is no doubt that humans have played their part in the history of the dunes which means that the dunes that we see and the infiltration rate might be influenced by people.

e **4/6 marks awarded.** The explanation about the fieldwork equipment is much stronger than that about human influence. The discussion of fieldwork equipment makes reference to both the influence on results and on conclusions, and scores the full 3 marks. The second factor is less well explained and needs further development to show how human interference might have affected both the results and the final conclusions.

(e) **Discuss the purpose of statistical analysis as part of any fieldwork study and explain briefly why your chosen statistical method was selected as suitable for your fieldwork.** (4 marks)

e Statistical analysis helps students to test their hypotheses in any investigation. It helps geographers to reach valid, reliable geographical conclusions. It allows a lot of data to be compared in a meaningful and concise manner. For this question, 2 marks are available for an awareness of the purpose of statistical analysis in any fieldwork investigation and 2 marks are for a justification of the chosen method. You need to make reference to aim/hypothesis and show why this technique was suitable.

(e) Statistical analysis in a fieldwork investigation provides a method of processing, calculating and interpreting raw data into a result which can help to prove or disprove the hypothesis. Spearman's rank was chosen to give a result, in this case, $r_s = 0.94$, which allows us to interpret the degree of correlation which was highly significant. This helps prove our hypothesis, in relation to infiltration rates, which stated 'The time taken for water to infiltrate will increase back through psammosere'. Our Spearman's rank showed that this was a suitable measure as the relationship between the two variables could be measured and the strength of relationship could be commented on.

e **4/4 marks awarded.** The first element of the question is answered well and some of the reasons why statistical analysis can be important are explained. The reasons why Spearman's rank is appropriate in this case study are explained and the outcome of using this technique is shown clearly and linked to the aim of the study. This is a well-argued answer that scores full marks.

Question 2 Geographical techniques

Two questions have been supplied in this section to demonstrate the different types of questions that you might encounter in the exam; each question is worth 30 marks.

Question 2A

(a) Study the table below, which relates to an investigation into the relationship between GDP per capita and literacy rate across 15 countries.

Country	GDP per capita (US dollars)	Rank	Literacy rate (%)	Rank	Difference in ranks (d)	d^2
Kuwait	45,455	1	94	5	–4	16
Ireland	43,592	2	99	2.5	–0.5	0.25
Finland	38,655	3	100	1	2	4
Saudi Arabia	31,729	4	86	9		
Argentina	12,034	5	98	4	–1	1
Brazil	11,909	6	90	6	0	0
Egypt	6,724	7	72	12	–5	25
Guatemala	5,102	8	69	13	–5	25
Samoa	4,517	9	99	2.5		
India	3,876	10	74	11	–1	1
Bangladesh	1,883	11	79	7	4	16
Kenya	1,766	12	87	8	4	16
Zambia	1,712	13	80	10	3	9
Burkina Faso	1,513	14	22	15	–1	1
Ethiopia	1,139	15	39	14	1	1
						$\sum d^2 = 182.5$

Complete the table by filling in the four missing values. (2 marks)

ℯ This is a straightforward task — 2 marks if all four are correct, 1 mark for two or three correct. Some students forget to put the minus sign in front of the 5 and lose a mark.

Student answer

(a)

–5	25
6.5	42.25

ℯ **2/2 marks awarded.** The marks are awarded for the completion of the correct figures. Accuracy is important and the minus sign MUST be present. If all four are correct — 2 marks. If 2 or 3 are correct — 1 mark only.

(b) Calculate the Spearman's rank correlation for the data shown and make a comment on the statistical significance of the result. (The formula and significance graph and table are found on pages 11–12.) *(6 marks)*

ⓔ There are 2 marks for getting the first breakdown of the formula correct and another mark for taking the formula through before the division. There is 1 mark for getting the right r_s value (to 2 significant figures). The final 2 marks are given for the comment on the significance. 1 mark is awarded for noting that this is a significant result and 1 mark for noting that this is found between the 95% and 99% levels.

(b) $r_s = 1 - \left(\dfrac{6\sum d^2}{n^3 - n}\right)$

$r_s = 1 - \left(\dfrac{6 \times 182.5}{15^3 - 15}\right)$

$r_s = 1 - \left(\dfrac{6 \times 182.5}{3{,}375 - 15}\right)$

$r_s = 1 - \left(\dfrac{1{,}095}{3{,}360}\right)$

$r_s = 1 - 0.33$

$r_s = 0.67$

The result is very significant and shows a strong positive correlation.

ⓔ **5/6 marks awarded.** The processing of the Spearman's rank correlation is carried out correctly but the answer does not go into enough depth when describing the statistical significance. You need to use the graph/table to work out that the answer is within the 95–99% arc. The question does not require comment on the correlation/relationship for the graph, so no marks are awarded for this.

(c) Explain some geographical reasons that could account for this particular statistical result. *(4 marks)*

ⓔ As the GDP per capita increases there is a strong positive correlation with the literacy rate. This calculation shows that there is a strong relationship between the two variables. The answer needs to explain this relationship — maybe countries that are richer will have more money available to provide better/free/accessible education for their citizens. 4 marks are awarded for a well-argued response. Statistical reasons are not credited.

(c) The table and the Spearman rank result show that the result is a positive correlation. This means that there is a link or a relationship between the two variables. As GDP increases this might have a positive impact on literacy. This is not surprising as governments in poor countries might not have much money to invest in the education system and this will have a big impact on literacy rates.

e **3/4 marks awarded.** This identifies the main relationship and goes on to explain a potential reason as to why this might be happening in this particular case. It could have gone into a little more detail to explain why this might be the case.

(d) Some geography students carried out a piece of fieldwork on the sand dunes at Portstewart Strand. They measured the amount of time that it took for 200 ml of water to completely infiltrate into the sand/soil across the sand dune. Their results are shown in the following table.

Distance from sea/m	0 (sea)	30	60	90	120	150	180
Infiltration rate/s	12	34	67	105	65	204	204

State the mode for the data in the table above. (1 marks)

(d) The mode is the most frequently occurring number in the data set, so 204/s.

e **1/1 mark awarded**

(e) Calculate the median for these data and state **one** disadvantage of using this statistic to summarise the data. (3 marks)

e The median is the central value when all of the values are ranked in order. In this case the median is 67 (the fourth number in the sequence).

(e) The median is 65. The median is not as reliable as other measures of standard deviation such as the mean because the median is not good when comparing small amounts of data, and it is difficult to use the median when going onto further statistical techniques.

e **2/3 marks awarded.** This shows good understanding and depth when discussing the problems/disadvantages of using the median in statistical analysis. However, it identifies the wrong value for the median so fails to score the mark.

(f) Study Figure 17 on page 25, which shows the percentage of total electricity production by generating source for selected countries.
The table below has been partially completed, illustrating the percentage contribution of hydroelectric, nuclear and thermal (and other) types of energy production. Use the triangular graph to complete the table. (4 marks)

Country	France	Sweden	USA	Canada
Thermal and other (%)	23			29
Hydroelectric (%)	25			59
Nuclear (%)	52			12

e This is fairly straightforward. 1 or 2 correct answers = 1 mark, 3 or 4 correct answers = 2 marks, 5 correct answers = 3 marks. Answers must be ±1% to be awarded a mark.

(f)

Sweden	USA
22	74
48	12
30	14

ⓔ **4/4 marks awarded.** All the values are correct.

(g) When might a triangular graph be an appropriate method of presentation? (2 marks)

ⓔ Triangular graphs can only be used in geographical studies when there are three separate variables and each component must be able to be measured out of 100%. There is 1 mark for identifying that only three variables are found and 1 mark for noting that we use percentages for this.

(g) A triangular graph is appropriate when you are investigating something that includes three parts, something simple and something that includes percentages.

ⓔ **2/2 marks awarded.** This identifies both the need for three parts and the percentages, although understanding appears to be weak.

(h) Study Figure 6 on page 18, which shows the world population density by country. Answer the questions that follow.
  (i) Identify the population density for the UK. (1 mark)
  (ii) Name the mapping technique used to show information like this. (1 mark)
  (iii) Discuss **one** advantage and **one** disadvantage of using this type of map to present geographical information. (6 marks)

(h) (i) 200–500 people per km^2

ⓔ **1/1 mark awarded.** There is 1 mark available for the right answer but the unit of measurement must be given.

(h) (ii) Choropleth

ⓔ **1/1 mark awarded.** 'Area-shaded map' is also acceptable.

(h) (iii) Choropleth maps are easy to draw and show data in a simple and straightforward manner. However, it is argued that they can oversimplify the patterns — areas that are densely populated will not actually show the variation within the area or country.

ⓔ 3 marks are available for a well-developed discussion of an advantage and 3 marks for a well-developed discussion of a disadvantage. Advantages might include that choropleth maps can be easy to read and can provide a good visual presentation of the data. Limitations might include the fact that sometimes the maps can be oversimplified so that large regions will only indicate one value, when the range within the area might be wide. Also, maps can sometimes provide big contrasts at borders that are unrepresentative of reality.

ⓔ **4/6 marks awarded.** This goes into some depth in relation to both the advantages and disadvantages. There is enough for 2 marks out of 3 for the first point. The disadvantage is well stated and the variety within the shaded area is noted. This point is maybe even better than the first point but it still lacks clarity and therefore only gets 2 out of the 3 marks.

Question 2B

(a) A geography student used nearest neighbour analysis to investigate the distribution of Met Office automatic synoptic and climate stations across Northern Ireland in 2012. The following hypothesis was proposed:

'The distribution of automatic weather stations across Northern Ireland is significantly random.'

Study Figure 20 on page 31 which shows a map of the Met Office automatic synoptic and climate stations in Northern Ireland in 2012. The table below is a partially completed version of the nearest neighbour analysis of the distribution.

Met Office station	Number	Nearest neighbour	Distance (km)
Giant's Causeway	1		20
Ballypatrick Forest	2	1	20
Portglenone	3	16	20
Killylane	4	6	12
Orlock Head	5	9	10
Aldergrove	6	4	12
Helens Bay	7		10
Stormont Castle	8	7	10
Ballywatticock	9	5	10
Murlough	10	12	20
Killowen	11	12	22
Katesbridge	12	10	20
Glenanne	13	12	25
Derrylin	14	15	22
St Angelo	15	14	22
Lough Fea	16	3	20
Castlederg	17	15	25
Banagher	18	19	20
Ballykelly	19		12
Magilligan	20	19	12

Map area = 165 km × 130 km = **22,275 km²**

$\sum d = 344$

Using Figure 20 on page 31, complete the table by filling in the missing values. (3 marks)

Student answer

(a) Giant's Causeway, 2; Helens Bay, 9; Ballykelly, 20.

ⓔ **3/3 marks awarded.** All the values are correct, for 1 mark each.

(b) Complete the nearest neighbour analysis (R_n calculation) and state the type of distribution shown in Figure 20. The nearest neighbour index equation and significance graphs are shown on pages 31 and 33, respectively. Comment on what this result indicates about the hypothesis stated. (6 marks)

ⓔ There are 4 marks available for the successful application of the whole formula and 2 marks for the comment on the hypothesis. The distribution of the weather stations is nearly perfectly random, with a slight tendency towards clustering. However, as only 20 sites were measured, this random result is considered to be at the 95% significance level, which means that we should reject the hypothesis because we do not have enough information to say that this is significant. 1 mark is for noting the random pattern and 1 mark for suggesting that the hypothesis would have to be rejected in this case.

(b) $R_n = 2d\left(\sqrt{\dfrac{n}{a}}\right)$

$R_n = 2 \times 17 \left(\sqrt{\dfrac{20}{22,275}}\right)$

$R_n = 34.4 \times \sqrt{0.0008}$

$R_n = 34.4 \times 0.028$

$R_n = 0.96$

The distribution is significantly random.

However, this hypothesis would have to be rejected as random is at the 95% level – not enough points have been used in this case.

ⓔ **6/6 marks awarded.** The formula has been used properly and the type of distribution is correctly identified. The link to the hypothesis is also correct. This is a well worked out answer.

(c) When using nearest neighbour analysis to identify a distribution pattern, a number of different factors can influence the final result of the R_n value. Describe and explain **one** factor that could affect R_n values. (3 marks)

ⓔ Most answers will involve a comment that discusses the influence of area on the R_n value. They should be able to effectively describe how a larger area will usually produce a more clustered result. However, in this case it is likely that they make reference to the number of points as well — as the points were randomly scattered and this does mean that a random distribution was a distinct possibility. Perhaps the area chosen for the map was too big in this particular case.

(c) This R_n value was greatly affected by the fact that there were not enough points available for the actual statistic to be used properly. Ideally, 30 sites are needed for the statistic to work properly. If 30 sites were used this might show that places are quite regular (and the R_n will increase) so that the whole of NI is covered.

ⓔ **3/3 marks awarded.** This answer includes some useful points and argues clearly how the R_n result might actually change, making a valid statement that describes and explains how the R_n could be affected.

(d) Study the table below. It shows the results of an investigation that measured the depth of the river at 30 cm intervals across a meander section of the Glenarm river in County Antrim.

Site	Distance from the inside bank (cm)	Depth of river (cm)
1	0	0
2	30	5
3	60	7
4	90	14
5	120	23
6	150	29
7	180	35
8	210	45
9	240	45
10	270	33
11	300	50
12	330	58
13	360	67
14	390	55
15	420	43

Using the data above, plot the information onto a graph and label it fully. (7 marks)

ⓔ 1 mark would be awarded for 1–3 sites labelled accurately, 2 marks for 4–8 sites labelled accurately, 3 marks for 9–13 sites labelled accurately, 4 marks for all sites labelled accurately and marked. 1 mark is for the title, 1 mark for labels on the x- and y-axes and 1 mark for making sure that line connections are accurate. Either a correlation or a line graph is acceptable.

(d)

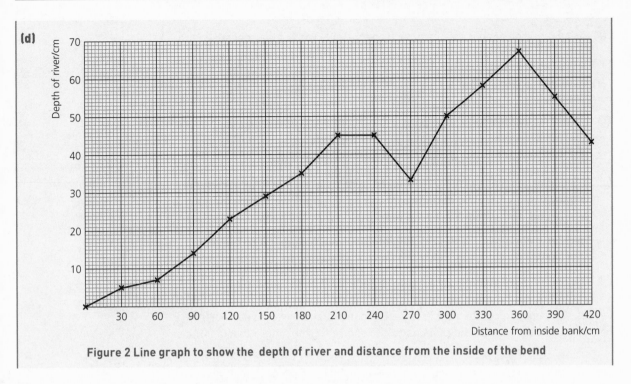

Figure 2 Line graph to show the depth of river and distance from the inside of the bend

ℯ 7/7 marks awarded. The title is clear, the axes are clearly labelled and all sites are clearly and accurately marked on the graph.

(e) Describe the pattern on your graph from (d). (3 marks)

ℯ You need to describe the relationship between the two variables on the graph. There must be some clear description to show that the depths increase towards the other side of the river. Figures must be used in order to score the maximum 3 marks.

(e) The graph shows that in the inside bend the depths are quite shallow (0 cm at the start and then 30 cm across the river the depth is only 5 cm). As you move across the river from one side to the other the depth increases though there is a patch in the middle of the river where the river bed flattens out and there might be the odd bigger stone sitting. The deepest part of the river (at 360 cm across the river) is 67 cm deep and the river starts to get shallower towards the far side.

ℯ 3/3 marks awarded. This answer includes some good detail and uses figures to show the main features and changes in the depth of the river. Good understanding is shown of both sides and aspects of the river meander.

(f) Suggest one geographical reason to explain this pattern. (3 marks)

ℯ The most common answers here will address the differences in velocity and discharge on the two sides of the river, and the link with energy and erosion. You need to make sure that you refer to the features on both sides of the river to get 3 marks.

(f) The river gets deeper as you move from one side to the other because of the process of erosion. Towards the outside of the river bend the deepest part of the river is also the place where the water moves the fastest and erosion here (in particular hydraulic action) will be strong.

🅮 **2/3 marks awarded.** This is a good answer that deals with the erosion aspects and how the deep sections of the river are formed, but it ignores the shallow sections where there might be evidence of deposition.

(g) A group of geography students is asked to complete a questionnaire survey to investigate where the parents of the school's pupils were born for each year group. The school has seven year groups with 100 people in each year group.

Identify and explain which sampling method you think might be the most appropriate for the geographers to use to carry out this questionnaire. (3 marks)

🅮 The main type of sampling that would be appropriate here is stratified sampling. The researchers would need to get an equal sample/ number of responses from each of the year groups. Random sampling would not work because this might group the responses into one year group more than another.

(g) You would need to get an equal sample of responses from each of the year groups so you would organise this so that you got the same number of questionnaires from each of the year groups in the school.

🅮 **1/3 marks awarded.** The student has not identified the type of sampling but then does go on to describe accurately how this might be carried out, although even this needs to be developed further.

(h) Suggest how many questionnaires you think the geographers might need to sample this particular case, and briefly explain your answer. (2 marks)

🅮 There is no definitive number of questionnaires that might be taken in a survey such as this — it depends on how accurate and representative the researcher wants to be. The 2 marks are for a well-argued and sensible approach. Generally, no less than 10% should be taken but it might be better to take a representative sample of 20% in this particular case. This means that between 10 and 20 questionnaires should be taken from each year group (with half going to boys and half to girls). This means that a minimum of 70 questionnaires might be needed.

(h) To get an accurate example of the information, each year group could get 50 questionnaires — with 25 for boys and 25 for girls. This would be 50 × 7 = a total of 350 surveys would be taken.

🅮 **2/2 marks awarded.** This answer addresses the question well and explains how many surveys might be taken and how this might be broken down.

A2 Unit 3

■ Decision-making question

This question will take a different form to the other questions in this guide. The information will be presented in a similar format to what you would expect to see in the exam paper and then the answers will be analysed at the end.

Read the background information below and write a report in response to the task that follows.

You are recommended to spend 30 minutes reading this section and 60 minutes writing your answers.

Background

The Tourism NI marketing team note that their main aim is to 'develop marketing and promotional platforms to showcase product, build itineraries, develop trails and enhance web presence targeted to key consumer groups'.

In recent years, Tourism NI have attempted to use the global success of local golfers Rory McIlroy, Graeme McDowell and Darren Clarke to promote Northern Ireland in relation to golf tourism. Tourism NI note that they 'have a role in product development, working with golf clubs and associated golf bodies and agencies to ensure the golfing visitor has a quality visitor experience throughout their trip'.

Key market data

- The economic value of golf tourism in Northern Ireland in 2014 was £33.2 million. In 2013 the economic impact was estimated at £27 million. There was a year-on-year increase of £6.2 million or 23%.
- The number of golfing visitors to Northern Ireland was 139,000 in 2014, up from 134,268 in 2013 — an increase of 3.7%.
- This means that the average economic impact per visitor in 2014 was £238.

Tourism NI have developed a **'Golf Strategy for Northern Ireland to 2020'** which was released in March 2015.

In 2012 one of the UK's biggest golf course development projects was announced for Northern Ireland. A £100 million, 365-acre golf resort was planned to be built on the Bushmills dunes near the Giant's Causeway on the County Antrim north coast. The plan included a five-star hotel, 70 golf lodges, a Golfing Academy and a championship-class 18-hole links golf course. The plan was estimated to bring an additional 360 jobs to the area.

A similar project was developed by Trump International Golf Links to the north of Aberdeen in Scotland and it became controversial in relation to the impact that it would have on the economic, social and environmental landscape.

The Bush Dunes project was initially led by the late local businessman Dr Alistair Hanna but met with both praise and opposition from a very early stage. Much of the opposition came as the proposals meant that it would be built right next to Northern Ireland's premier tourist attraction, the UNESCO World Heritage Site and National Nature Reserve, the Giant's Causeway.

Although plans have stalled for this development another local businessman has taken an interest in the site and is keen to pursue the Bush Dunes golf course project.

Task

You must base your answer solely on information contained in this examination paper and resource booklet and not on any other information or decision that might be available in relation to this issue.

You must adopt the role of the Northern Irish Minister for the Environment, who is directly responsible for large planning considerations. You need to consider whether this golf course development is economically and environmentally viable and whether or not it should proceed.

Structure for answer

	Marks	
Format	2	Each of the three sections needs to be set out using the headings provided.
Role	2	You must take on the stated role.
Graph	8	Draw a graph using the data for 2014 from Resource D and make reference to it in the report at an appropriate place.

Your report must be structured as below		Marks		
Heading	Sub-heading	Sub-section	Section	Guidance to candidates
A Introduction		10	10	**A** Explain the reasons why a massive golf project like this is being considered in Northern Ireland and briefly outline the proposed project.
B The likely impact on:	(i) economy	10	28	**B (i)** Discuss the possible beneficial effects the proposed development might bring to the economy in the Causeway Coast area, and the counterarguments.
	(ii) environment	10		**B (ii)** Discuss the potential environmental damage of the proposed development and the counterarguments.
	(iii) social	8		**B (iii)** Discuss the potential social consequences (especially on the local people) of the proposed development, and the counterarguments.
C Decision		10	10	**C** State clearly your decision and justify it on the basis of the greater overall benefits.

Checklist of resource material

Maps

Resource A   OS map of the Bushmills to Giant's Causeway area

Resource B   Map of the area

Images

Resource C1   Looking across the proposed area for the new development
                  from Bushfoot Golf Course

Resource C2   The Giant's Causeway

Resource C3   The Giant's Causeway viewed from the top path

Resource C4   Looking across the proposed area for the new development
                  from the Causeway coastal path

Resource C5   The Causeway coastal path looking towards Runkerry House

Resource C6   Looking across the proposed area from Runkerry House

Resource C7   Plan of the resort development

Resource D   Top ten visitor attractions 2011–2014 in Northern Ireland

Text

Resource E   Introduction to the proposal for the Bush Dunes development

Resource F   Arguments in favour of the proposal

Resource G   Arguments against the proposal

Resource H   Quotations related to the development

Resource A

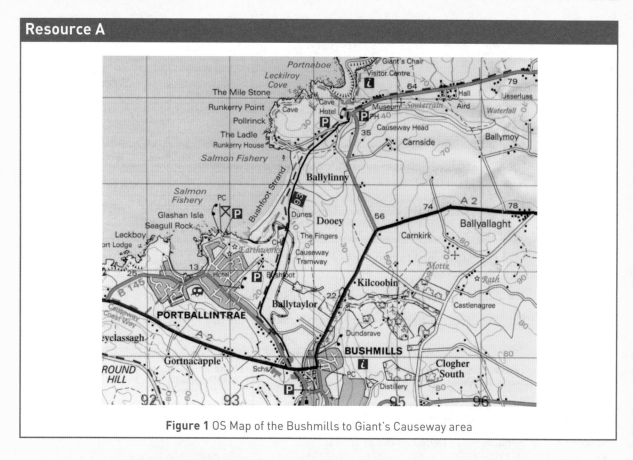

Figure 1 OS Map of the Bushmills to Giant's Causeway area

Resource B

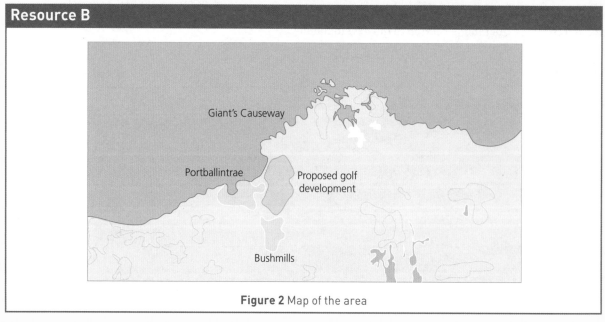

Figure 2 Map of the area

Resource C1

Figure 3 Looking across the proposed area for the new development from Bushfoot Golf Course

Resource C2

Figure 4 The Giant's Causeway

Resource C3

Figure 5 The Giant's Causeway viewed from the top path

Resource C4

Figure 6 Looking across the proposed area for the new development from the Causeway coastal path

Resource C5

Figure 7 The Causeway coastal path looking towards Runkerry House

Resource C6

Figure 8 Looking across the proposed area from Runkerry House

Resource C7

Figure 9 Plan of the resort development

Resource D

Table 1 Top ten visitor attractions 2011–2014 in Northern Ireland

	Attraction	Number of visitors (thousands)				Change 2013–2014 (%)
		2011	2012	2013	2014	
1	Giant's Causeway World Heritage Site	533	524	754	788	5
2	Titanic Belfast	n/a	665	604	634	5
3	Ulster Museum	471	595	416	466	12
4	Derry's Walls	278	281	411	370	–10
5	W5 (Who, What, Where, When, Why)	251	328	297	324	9
6	Carrick-A -Rede Rope Bridge	243	276	263	323	23
7	The Guildhall	35	n/a	269	299	11
8	Belfast Zoo	283	259	236	253	7
9	Pickie Fun Park	n/a	277	266	243	–9
10	Murlough National Nature Reserve	n/a	n/a	358	241	–33

Resource E

Introduction to the proposal for the Bush Dunes development

Source: www.breakingnews.ie/ireland/pledge-to-protect-world-heritage-site-as-golf-course-gets-go-ahead-540675.html#ixzz1ngupJunF

The first planning application for a resort development like this was submitted in 2001 and renewed six years later. Dr Alistair Hanna said:

'This is a unique project which will be world class in every aspect. The dunes are phenomenal. Every course architect who inspected the landscape has raved about the place. They've said: "The piece of earth is just made for golf." It's amazing. There just isn't anywhere else like it in the world. With Royal Portrush, Portstewart and Castlerock (golf clubs) in the same area, I want this part of the world to become a gold resort on a par with Pinehurst and Pebble Beach (in the USA). I know this is a difficult time economically, but times will get better. We are not building for today. We are building for tomorrow. Golf in 2020 will be in a different place from where it is today and I want this place to be among the top 10 golf destinations in the world. There is no better time than now, especially in terms of developing Northern Ireland's golfing profile.'

Complex designer Richard Hunter, from nearby Ballymoney, insisted that the development was 'environmentally sensitive.' He said the built-up parts of the resort were some distance from the causeway. 'We have designed the buildings in such a way that they create the lowest possible visible impact,' he added. 'There is no part of it more than three storeys, it's been developed along the side of an existing slope so we've designed it in such a way that it blends in very sensitively with the existing landscape, we've used local stone for walling generally and the roofs will be grassed so it will create a low level sweep.'

The Department of the Environment (Northern Ireland) note the following about the Giant's Causeway and Causeway Coast (UK)

'The Giant's Causeway and Causeway Coast World Heritage Site (WHS) was inscribed as a natural site in 1986 on the basis of its demonstration of two outstanding universal values summarised as geological phenomena and natural beauty. The features of geological and geomorphological interest which are globally significant are fully represented within the property. While the natural beauty is related to the cliff exposures of columnar and massive basalt within the site, this is set within more extensive coastal scenery.'

Resource F

Arguments in favour of the proposal

A strategic review of golf tourism in Northern Ireland 2015–2020

Adapted from: www.tourismni.com

This strategy sits in the context of Northern Ireland's increased profile as a golf destination, the unprecedented success of our home-grown golfers and our strong performance in delivering and attracting major events. All this has resulted in an upsurge of demand from the golf consumer, travel press and tour operators and political

recognition that golf has the ability to contribute to the economic well-being of Northern Ireland.

Globally, golf tourism is estimated as a $32 billion per annum market. Golf operates on a huge scale. There are estimated to be some 60 million golfers worldwide who play around 30,000 courses across the planet. The worldwide market for international golf travel is said to be 'buoyant'. Global golf holiday sales grew by 9.2% in 2014 (following 9.3% in 2012 and 11.1% in 2013). This increase came at a time of economic downturn.

Golf tourism is estimated to generate £220 million in Scotland per year. In the Republic of Ireland an estimated 163,000 overseas golfers visiting in 2012 generated around £202 million. Recent research also suggests that golfers are amongst the most valuable visitors to Ireland, spending around £1,200 per person on each trip — more than 2.5 times that of the average overseas visitor.

The Golf Tourism Strategy is designed to support the golf sector contribution to the wider tourism vision of a £1 billion industry in Northern Ireland by 2020. Golf tourism in Northern Ireland is currently valued at some £33 million so in terms of Northern Ireland's competitiveness — vis à vis other golf destinations — there are real opportunities to further grow golf tourism.

The aims of the Strategy are to:
1 Grow the value of golf visitors to Northern Ireland to £50 million p.a. by 2020
2 Develop the capability and capacity of clubs to host visitors
3 Enhance the reputation and visibility of Northern Ireland for golf tourism in designated markets
4 Build Northern Ireland's capacity to host golf events
5 Support industry leadership and collaboration across the sector.

Extensive stakeholder engagement and consultations have yielded a fresh approach and this Strategy provides a cohesive rationale, action plan and identification of roles and responsibilities within the industry to deliver on these objectives.

There are 69 18-hole golf courses in Northern Ireland with the premier courses of Royal County Down and Royal Portrush being a major reason for more wealthy tourists to visit.

In recent years, Tourism Northern Ireland has attempted to position Northern Ireland's profile in the golfing marketplace by using the single uniform message: 'Northern Ireland Made for Golf'. The idea was that this clear, straightforward message would help create a strong recognisable identity that would promote a sense of place for golf in Northern Ireland (i.e. links to positive aspects of golf, the natural landscape, the Northern Ireland 19th hole experience, etc.).

The potential economic benefit of a variety of Golf Tournaments

	The NI Open	The Irish Open (2012 at Portrush)	The Open Championship (coming to Portrush in 2019)
Tour	PGA Challenge Tour	European Tour	Major Championship
Visitors	34,000	132,000	200,000 (est)
Bed nights	6,000	28,000	50,000 (est)
Economic impact	£1 million	£12 million	£80 million (est)

Northern Ireland Annual Tourism Statistics 2014 (NISRA)

Source: www.detini.gov.uk/sites/default/files/publications/deti/2014-Annual-publication.pdf

The most recent Northern Ireland census of employment figures indicates a 3% increase in tourism-related jobs between 2011 and 2013. This increase was mainly in 'accommodation for visitors' and 'food and beverage serving' industries. Around one in ten of all employee jobs are in tourism-related industries.

Overall in Northern Ireland in 2014:
- There were 4.5 million overnight trips by all visitors (up 11% on 2013).
- Associated expenditure increased by 4% (to £751 million in 2014).
- External (outside NI) overnight trips increased by 4% to 2.2 million.
- More than 15 million visits were made to local visitor attractions, a 3% increase on 2013. The top attractions were the Giant's Causeway and Titanic Belfast.

- Hotel room occupancy stood at 65%, up from 64% in 2013. In total 1.85 million hotel room nights were sold in 2014.

- Last year 69 cruise ships docked in Northern Ireland with up to 121,000 passengers/crew on board. The number of cruise ships docking here has doubled over the last three years.

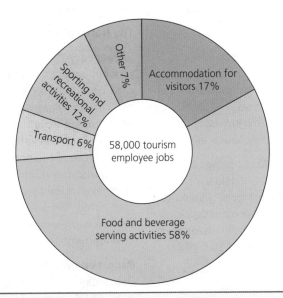

Pie chart: 58,000 tourism employee jobs
- Accommodation for visitors 17%
- Food and beverage serving activities 58%
- Transport 6%
- Sporting and recreational activities 12%
- Other 7%

Resource G

Arguments against the proposal

Source: www.facebook.com/GreenParty/posts/859981100689128?fref=nf

The Leader of the Green Party in Northern Ireland, Mr Steven Agnew, noted:

'The North Coast is one of our greatest natural assets and should be afforded the utmost protections. It is of course the location of the Giant's Causeway which is our only UNESCO World Heritage Site and this proposed golf resort development poses a real risk to that prestigious ranking which attracts over half a million visitors each year.

This area is one of the key attractions for local, national and international visitors to Northern Ireland, as well as an important protected area of nature and it should not be open to exploitation which risks its wide appeal. Last year a report carried out by a UNESCO advisory body stated that the golf course would 'create an irreversible change of landscape character' in a protected area of outstanding natural beauty.

Neither UNESCO nor the National Trust supported this development because both organisations feared the impact the golf course would have on this very special area.'

Friends of the Earth are another group to voice opposition. James Orr, the NI director, said:

'Many discerning tourists will not expect to see a new hotel, practice greens, lighting, new houses and manicured lawns so close to the spectacular wildness of the Giant's Causeway.

A sensible approach to managing our economy would protect the assets of our most important tourist attraction – the Giant's Causeway. There is a sustainable supply of jobs in conservation and heritage. Why put this at risk?

Planning policies for that area are very clear – the landscape around the Giant's Causeway should be protected. Instead, a form of landscape trauma is being permitted at Northern Ireland's only World Heritage Site. It's like building a drive-through burger bar at the Taj Mahal.'

ASSI

The proposed site for the development is next to Bushfoot Beach and forms part of Runkerry ASSI (Area of Special Scientific Interest). The Runkerry ASSI notes that this place is a 1.2-km-long strand. It is said to be the highest energy beach in Ireland.

'A beach system of international importance demonstrating beach states from dissipative to reflective. A wide range of rhythmic morphological features are present including beach cusps, rip current channels, longshore rip feeder channels, giant cusps and migratory transverse and crescentic nearshore bars.'

The NO Runkerry Golf Complex Group

A social media campaign using Blog entries on Wordpress and through Facebook was set up by Dick Glasgow in opposition to the proposed golf development. Mr Glasgow in particular was most concerned at the impact this development might have on the local animal life. For example, he noted:

'Pesticides are applied to golf courses at higher concentrations per acre than almost any other types of land, including farmland, and there are concerns that their extensive use could contaminate waterways and damage neighbouring communities and wildlife.'

He also notes the letter that David McNeill, the Botanical Society of Britain and Ireland (BSBI) recorder for County Antrim, wrote to the Planning Office in relation to this proposal. He was aware of nine very rare flower species which grow in the site. Four of these have Northern Ireland priority species designation: Spring Vetch, Scots Lovage, Field Gentian and Chaffweed. The five other very rare species are listed as: All-seed, Early Forget-me-not, Common Cudweed, Sea Bindweed and Bugloss.

In relation to animals, the Bushfoot dunes are acknowledged as being one of the last remaining habitats for the very rare Mining Bee. In addition, some rare bats are found nesting in an area that would be removed for the development.

The National Trust

In February 2012, the National Trust said 'it was not opposed to the development' but was convinced that 'the planning application was contrary to a range of the Department for the Environment's planning policies'. A spokesperson said: 'As a conservation charity the trust's over-riding focus is the protection of the environment and landscape within the distinctive setting of Northern Ireland's only World Heritage Site.' They also noted, 'We believe this farmland and dune system is the wrong place for such a massive development.'

Report on the Advisory Mission to the Giant's Causeway and Causeway Coast (February 2013) by UNESCO (designators of World Heritage Site status)

A letter from the DoE Northern Ireland to the World Heritage committee in 2012 noted that:

'It is acknowledged that the nature of the proposal is such that it will have a significant landscape and visual impact on the setting of the Giant's Causeway and Causeway Coast World Heritage property, and the Causeway Coast AONB.'

The UNESCO report also notes the different Protect Area Legislation that is in force in this area:
- Giant's Causeway National Nature Reserve (primarily to recognise the diverse/important plant communities)
- Giant's Causeway and Dunseverick Area of Special Scientific Interest (which covers the beach system)
- Causeway Coast Area of Outstanding Natural Beauty (which recognises the quality of the landscape, its scenic quality and the diversity of the coastal landscape from Portrush to Ballycastle)

- North Antrim Coast Special Area of Conservation
- The Skerries and Causeway is a candidate Special Area of Conservation site

A major new development, called the Runkerry Development, has been proposed to sit around 550 metres south of the World Heritage property and within the 'distinctive landscape setting' close to the World Heritage Site. The core of the development site, 'of approximately 148 hectares', is located about 2 km to the south of the Giant's Causeway visitor centre. This would be double the size of the World Heritage Site area.

It has been acknowledged that the nature of the proposal is such that it will have a significant impact on the landscape and visual setting of the World Heritage Site and the AONB.

The UNESCO report notes that some damage to the landscape setting is expected due to inappropriate development or land use. The buildings and golf course will be clearly visible in views from the World Heritage Site and will be significant elements in the landscape. Their impacts on the visual and landscape values of the property include:

- 120-bedroom hotel: although planned for the side of a hill and partly covered with grass roofs, the large amount of glazing on the building and the size of the complex of buildings will impact on visual and landscape qualities of the area.
- Other major buildings (clubhouse, golf academy, 3-hole practice facility) will be linked through a network of roadways, footpaths, street lightings, etc., and will create a visual impact.

- The 18-hole golf course will make a significant change in the landscape, through the introduction of manicured elements of the course and associated artificial infrastructure.
- 75 lodges would blur the transition between the settlement of Bushmills and the rural landscape, a key element in the experience of approaching the World Heritage Site.

The UNESCO report goes on to note that the attrition of natural features through natural processes such as cliff erosion or sea level change continues to be an issue. They are concerned about a gradual weakening of the underlying geology and undermining of cliffs by marine erosion and human excavations to facilitate the construction and maintenance of footpaths.

The report concludes that the proposed gold development constitutes a threat to the integrity of the property and its Outstanding Universal Value through an irreversible new landscape and visual impacts that affect the setting of the property, as well as the damage to the biodiversity which gives the wider landscape its character.

Amongst the 16 final recommendations from the UNESCO reporting team in relation to this issue, the first stated that:

R1: The impacts of the proposed development on the Outstanding Universal Value of the World Heritage property of Giant's Causeway and Causeway Coast appear sufficiently significant that the development of the golf course should not be permitted in its proposed location.

Resource H

Quotations related to the Development

Dr Alistair Hanna, Runkerry Developments CEO

'Not only will the resort provide a world-class golf links course and facilities attracting thousands of visitors each year, it will also protect the vulnerable topography of the coastal areas which has been left vulnerable following decades of neglect. This will be one of the most spectacular golf developments ever seen in Ireland.'

Source: http://www.bbc.co.uk/news/uk-northern-ireland-21607970

Heather Thompson, CEO The National Trust NI

'We believe that if a development of this scale does go ahead in this location, the message is that nowhere in Northern Ireland, no matter how important or protected, is safe from development. We have serious concerns for those partners involved in the care and protection of the world heritage site.'

Ian Paisley MP, North Antrim

Source: http://www.bbc.co.uk/news/mobile/uk-northern-ireland-17099679

Mr Paisley noted that he felt it was right that the development goes ahead.

'Their [the National Trust] actions I still believe have been disgraceful and damaging for the Northern Ireland economy, but we must . . . look to move forward as the course progresses.'

He also noted on another occasion:

'The Bushmills dunes course will be a significant asset for the region and a significant tourist facility. It will become one of the most iconic golf courses in the world, generate employment and write a new chapter in the history of this ancient coastline.'

Howard Hastings, Chairman of the Northern Ireland Tourist Board

Source: http://www.bbc.co.uk/news/uk-northern-ireland-22645799

'Runkerry is the right project at the right time for tourism and everyone should get behind it. The Northern Ireland Tourist Board has made it clear all along that the opening of a world class links resort so close to Royal Portrush and the other fantastic courses of the north coast will significantly boost our reputation as a gold destination in all our key markets and will encourage golfers from all over to stay longer and spend more time in Northern Ireland.'

A Introduction (10 marks)

As the Northern Irish Minister for the Environment, it is my job to make the final decisions on large scale planning decisions that will impact the people of Northern Ireland. It is my aim to consider the many different economic, environmental and social factors in this proposed development and come up with a final decision about whether it should proceed.

Tourism is a vastly important enterprise for the people of NI. Over recent years the numbers of tourists coming to our top five sites (Resource D) has gone up from 1.7 million visits to 2.5 million visits over three years. The most visited attraction (as can be noted from Graph A) is the Giant's Causeway World Heritage Site. The number of visits went up from 533,000 in 2011 to 788,000 in 2014. Tourism along the north coast is big business.

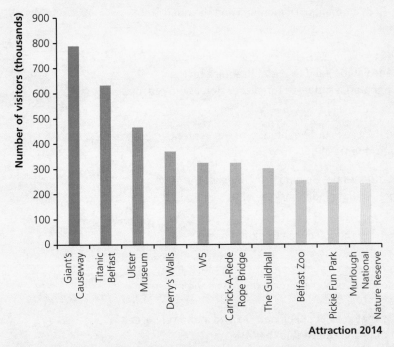

Graph A Bar chart of the top ten visitor attractions for Northern Ireland 2014

Over 58,000 people are employed in tourism-related industries throughout NI (Resource F) and the associated expenditure is £751 million for 2014. This money is vital to the success of the NI economy.

Golf tourism has become an important part of this. The success of local golfers like Darren Clarke, Rory McIlroy and Graeme McDowell has meant that people around the world are more interested in NI golf than at any other time. The value of golf tourism in 2014 was £33.2 million to the NI economy and the number of golf visitors increased to 139,000.

The development of the Bush Dunes project was to be part of the expansion of the NI Golf Strategy to 2020. The project is to be built between Bushmills and the Giant's Causeway, next to the existing Bushfoot Golf Course. The proposed golf resort site is a £100 million development which would cover a 365-acre site that would include a 5-star hotel, 70 golf lodges, a golf academy and a championship 18-hole golf course. The plans for the development were led by a local businessman, Dr Alistair Hanna.

However, the sticking point in relation to the project is that the site proposed for this golf course development is right on the doorstep of NI's only UNESCO World Heritage Site and Natural Reserve, the Giant's Causeway.

(e) These answers are all marked with levelled marking.

Level 3 (8–10 marks) will be accessible for an answer that clearly describes the proposed development and explains why a golf project like this might be considered for Northern Ireland. Both elements are included and the quality of written communication is excellent.

Level 2 (4–7 marks) will be awarded for answers that make fewer clear and correct points. Points that are made will be valid but might require further development. There might be an imbalance between the discussion of the reasons and the outline of the project.

Level 1 (1–3 marks) can be gained if there is little content or if the content presented is irrelevant to the requirements of the answer.

It should also be noted that CCEA offer some 'Guiding principles' at the start of their mark schemes for this section that guide markers of this exam to look favourably on answers that:

(a) Avoid undue verbatim quoting from the Resource Booklet and adopt a consistent style.
(b) Use the full range of the resource material appropriate to the task – particularly where it is provided in non-literary format such as the map and photographs.
(c) Apply knowledge and concepts that are not specifically raised in the resource material, yet are both illuminating and relevant to the task.
(d) Maximise opportunities presented by the resource material.
(e) Appreciate that 'bias' might exist in resource material which expresses particular views.
(f) Avoid undue repetition of the same answer material in different sections or, if overlap is unavoidable, present it in a fresh way.
(g) Back up points with specific detail, e.g. giving statistical information where it is provided rather than making vague statements when details are readily available.

(e) **9/10 marks awarded.** This is a strong answer that gets into the high section of the Level 3 mark band. The answer has clearly set the context for the golf project in Northern Ireland and has identified the wider tourism issues and hinted at how this will be placed within a Golf Tourism Strategy for Northern Ireland. There is also a brief outline of the key features of the project as required in the question.

ⓔ The graph has also been inserted as part of the answer to this question. The graph is worth 8 marks in this paper. Students need to think carefully before selecting the type of graph to be used. The most common graphs are bar charts, histograms and line graphs though scattergraphs and pie charts are also possible. You need to think about where is the best place to insert your graph and there needs to be some reference to the graph in the body of your written answer.

Make sure that the graph is drawn neatly and follows the usual conventions for the submission of a graph. Ensure that every graph:

- has a clear and simple title
- has both the vertical and horizontal axes clearly labelled (including any reference to units of measurement)
- has a key, when required.

ⓔ **8/8 marks awarded for graph.** The graph has been well presented. There is clear reference to the graph in the answer, the technique is appropriate for this type of data. Accuracy is excellent and proper conventions have been followed.

Student answer

B (i) The likely impact on the economy   (10 marks)

There is little doubt that the economic benefits linked to this development proposal outweigh any problems. As noted in my introduction above, tourism plays a very big part in the economic development of Northern Ireland. Tourism NI are keen that tourism becomes a £1 billion industry in NI by 2020.

Golf tourism in NI has a very important part to play if this target is to be realised. Golf tourism is a huge global market (worth over $32 billion per annum). The value of golf tourism in NI at present is valued at £33 million with a target to rise to £50 million by 2020.

Already there are 69 18-hole golf courses in Northern Ireland and the aim is to encourage more people to come. Resource F shows that there are a lot more people staying in NI hotels each year. Hotel occupancy is at 65% with 1.89 million hotel room nights sold in 2014. The development of gold resorts like the Bush Dunes will increase this further. For example, the Irish Open in 2012 in Portrush brought an additional 132,000 visitors, an additional 28,000 bed nights and an economic impact of £12 million. There are currently plans for the Open Championship to come to Portrush in summer 2019 and the estimated impact of this would be 50,000 bed nights and an estimated boost to the NI economy of £80 million.

It is hoped that developments like Bush Dunes would further increase the potential for tourism in this area. The investment brings £100 million into the local area and is planned to create 360 jobs.

However, the development is being planned at a bad time, Dr Hanna notes (in Resource E): 'I know this is a difficult time economically, but times will get better'. Developments like this will always put pressure on the local resources and local infrastructure. The building and success of the development could bring congestion to the local area and could have a wider impact on the number of people playing at other smaller golf clubs like the neighbouring Bushfoot golf course.

e The key question when looking at economic impacts is to consider the amount of money that will be brought into the area by such a development. How many jobs will be created? What impact will this have on the livelihoods of the people who live and work in the area? Is there a wider cost or pressure that will be put on the area and the people? Is there an economic price on any damage that could happen?

Level 3 (8–10 marks) suggested ideas will require a certain level of sophistication and need to show a clear understanding of the economic concepts. There needs to be some use made of the most relevant resource material and figures should be used to support the argument.

Level 2 (4–7 marks) will be available for answers where there is a more limited discussion and where there might be a lack of depth. There might be a heavy imbalance between the two sides of the argument.

Level 1 (1–3 marks) is reserved for answers that lean too heavily on copying the resources and those that have little understanding of the issues.

e **7/10 marks awarded.** The argument that sets out the benefits of the proposal is much stronger than the details given in the counterargument. This could have been developed further. However, there is a very good use and reference to the resources in the benefits section.

B (ii) The likely impact on the environment (10 marks)

The designer of the proposed Bush Dunes project, Richard Hunter (Resource E), has noted that he has tried to create an 'environmentally sensitive design' with the main built-up parts of the resort going to be 'some distance from the Causeway'. Buildings have been designed to have the lowest visible blending with the landscape – using local stone and the roofs are going to be grassed. Every possible measure will be taken to minimise the impact that the golf course and buildings will actually have on the local area.

There is no doubt that a golf course would help manage the area carefully and would bring new revenue and spending to what is currently a neglected part of the north coast (judging from the photographs).

However, there are many concerns about the negative environmental impact that the proposed development would bring. UNESCO in 2012 have noted that 'the nature of the proposal is such that it will have a significant landscape and visual impact on the setting of the Giant's Causeway'. It goes on to note that this will impact a unique beach system, a delicate plant community and an area of conservation that contains some rare insects. However, its main issue is one with the potential footprint of the buildings being considered.

The Green Party note than neither UNESCO nor the National Trust are in support of the development. Friends of the Earth go further and argue that 'many discerning tourists will not expect to see a new hotel, practice greens, lights, new houses and manicured lawns so close to the spectacular wilderness of the Giant's Causeway'.

Much of the proposed area has already been designated with Area of Special Scientific Interest (ASSI) status. The beach system is described as having 'international importance'. The wider environmental and wildlife issues are also raised by the 'NO Runkerry Golf Complex Group'. Dick Glasgow notes in particular the impact that this development could potentially have on wildlife: 'pesticides . . . could contaminate waterways and damage neighbouring communities and wildlife'. He also notes that nine very rare flower species are unique to this area and they have already been protected by the NI priority species designation. The National Trust note that 'we believe this farmland and dune system is the wrong place for such a massive development'.

ⓔ Most answers in a decision-making paper will include at least one opportunity to write about an aspect of the environmental impact. Every proposed development will have some level of impact but answers should deal with to what extent these are part of the planning and will be minimised. Sometimes issues will deal with pollution (visual, water or air) or will link to particular threats to habitats of plants and animals. However, sometimes the development will actually make improvements to an area (e.g. brownfield sites).

Level 3 answers (7–10 marks) will contain a certain level of sophistication and need to show a clear understanding of the economic concepts. There needs to be some use made of the most relevant resource material and figures should be used to support the argument.

Level 2 (4–7 marks) will be available for answers where there is a more limited discussion and where there might be a lack of depth. There might be a heavy imbalance between the two sides of the argument.

Level 1 (1–3 marks) is reserved for answers that lean too heavily on copying the resources and those that have little understanding of the issues.

ⓔ **8/10 marks awarded.** There is some very good use of the resource to demonstrate the main issues that a wide variety of interest groups have with this particular development, though these might have been taken a little further. What is the point of having planning legislation such as ASSI status if it is going to be ignored when a big investment is suggested? This is one of the most protected parts of Ireland and is on the doorstep of NI's *only* World Heritage Site, which means that it has international protection and any condemnation from UNESCO should be treated with the utmost seriousness. There is some reference to the counterargument and what the architect would have done to minimise any visual impact.

Student answer

B (iii) The likely impact on social (8 marks)

The social impact on the local people is not an easy thing to measure. The lure of 360 new jobs into the area is certainly one that would be positive for the people. This would allow young people the opportunity of a job when they leave school which means that they would not have to leave the area or go down to Belfast for employment. This means that the ageing population will not happen and services will remain all year round for the local area.

The development of this resource will put golf tourism in Bushmills on the international map. It could mean a new revenue stream with more tourism money and investment than ever before.

However, the resources clearly show that already there are concerns from the local people as to how this will be managed. There could be conflict amongst the local people as they come to terms with the balance between destroying a protected fragile environment and the desire for jobs and money.

ⓔ There is less obvious social impact material through the decision-making resources than what is available for the social and environmental issues so the candidate has to work a little harder to find impacts and to explain them.

Level 3 answers (7–8 marks) will contain a certain level of sophistication and need to show a clear understanding of the social concepts. There needs to be some use made of the most relevant resource material and figures should be used to support the argument.

Level 2 (4–7 marks) will be available for answers where there is a more limited discussion and where there might be a lack of depth. There might be a heavy imbalance between the two sides of the argument.

Level 1 (1–3 marks) is reserved for answers that lean too heavily on copying the resources and those that have little understanding of the issues.

ⓔ **4/8 marks awarded.** This is a more basic answer that just about struggles into level 2, though it sits towards the bottom of this mark range. Often there will be some idea of local reaction to any development in the quotations section. People need to consider the amount of conflict or opposition that there is behind one aspect of a proposal – it is often local people who feel that they are the custodians of local features. Sometimes, though, they also might consider how a development might improvement the quality of life in an area – will more facilities be provided? Will the local infrastructure be improved? Will it lead to more people living and staying in the area or will people be keen to move away?

C Decision (10 marks)

Having balanced all of the arguments set before me as the Minister for the Environment, it is my decision to accept the proposal and to allow the Bush Dunes Golf Resort Development to go ahead.

There are many positive reasons why this development should go ahead.

- Economically, the development of this resort is going to make the Bushmills area of vital economic importance internationally. People will travel great distances to play and stay at world-class facilities. The potential to increase earnings through golf tourism will be massive to the NI economy.

- This will provide jobs in the local economy but also in the wider tourism sector. This will also enhance the reputation of golf tourism in NI and will encourage some of the 60 million golfers around the globe to consider visiting the resort. The recent Irish Open and the Open Championship (scheduled for 2019) will further enhance the reputation for NI as a centre of excellence for golf. This is further developed through the planned marketing strategy which will firmly place NI on the global golf map. The impact on tourism in NI will be immense.

- This will continue to encourage rich travellers (and cruise liners) to continue to stop and feature NI on their list of ports to call.

- The development will improve a piece of land that is lying empty and is being used for nothing. This will encourage the local economy in Bushmills and will encourage people to come to visit other attractions in the local area when they are here. Tourism is one of the fastest growing economic sectors in NI and developments like this will further enhance the NI tourism message. Yes, there are some environmental issues but these can be minimised through some careful planning.

There is no doubt that there are some drawbacks and limitations to this development. These must be acknowledged and need to be addressed sensitively within the plans to develop this site.

- Environmentally: concerns raised by UNESCO in relation to the impact on the site of the Giants' Causeway World Heritage Site need to be addressed fully. Care needs to be taken to ensure that specific details on the placement and design of buildings in the area are agreed upon.

- Local animal habitats and vegetation need to be treated with sensitivity and should be identified, protected and transplanted to alternative local sites as appropriate. Local wildlife expertise should be brought in to liaise with builders as the site develops.

- Environmentally-friendly grass management systems need to be adopted for the golf course to ensure that pesticide and fertiliser use is kept to a minimum.

- Local people should be considered as decisions are made in relation to access to the site and its development.

I urge further safeguards to be built in to minimise any damage that might be made by this development and encourage the development company to work with local people and local environmental organisations to ensure that the impact of this proposal is reduced.

(e) The final decision needs to be stated clearly and a range of reasons given. Evidence for the arguments from both sides is required for a Level 3 answer (8–10 marks). There might even be further links made to the resource material. A clear grasp of the issue and the concepts raised is noted and points are relevant and logically structured.

Level 2 (4–7 marks) will be available for answers where there is a more limited discussion and where there might be a lack of depth. There might be a heavy imbalance between the two sides of the argument.

Level 1 (1–3 marks) is reserved for answers that lean too heavily on copying the resources and those that have little understanding of the issues.

(e) **8/10 marks awarded.** This is a good final section of the report. The candidate has clearly developed a strong sense of the role that they are expected to take. They have made a valid decision and have justified this with a variety of material taken from across the resources. There is a balance to the argument and the two sides of the argument.

Format (2 marks)

Role (2 marks)

(e) 2 marks are given for the student to show that any answer has followed the structure suggested on the question paper. This should usually be done by clearly labelling each section using the same titles as given in the 'Structure for Answer' section at the start.

The Role mark is given when the candidate clearly shows that they have assumed the role as noted in the exam question. In this case there is a clear reference to taking on the role of the Minister for the Environment in the first paragraph and then later again in a second place, in the final decision. Usually you will need to show that you have maintained your role in two separate places for the full 2 marks.

(e) **2/2 marks awarded for format.** Format has been used properly throughout the piece.

(e) **2/2 marks awarded for role.** Role has been used on two separate occasions through the report.

Knowledge check answers

1 Because primary data are recorded by the person doing the fieldwork, it is easier to assess how accurate and useful the data and results will be. If fieldwork is planned carefully, the primary data can make a big difference in the interpretation of results. However, collecting primary data can be time consuming and expensive (especially if specialist equipment is required) and can sometimes not be as accurate as it might be.

2

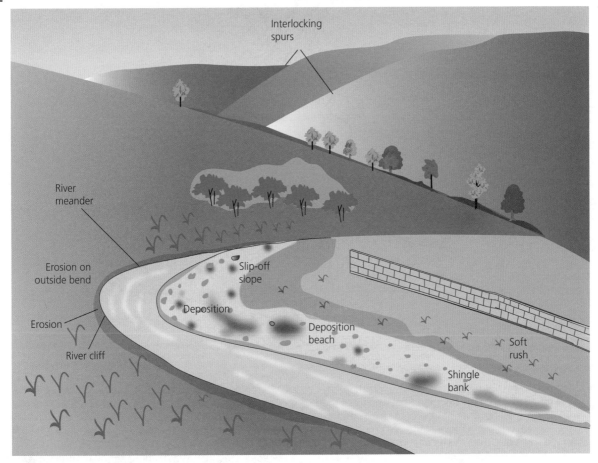

3 (a) Section D
  (b) Eagley Way and Darwen Road
4 (a) 6 and 12
  (b) 5

5 (a)

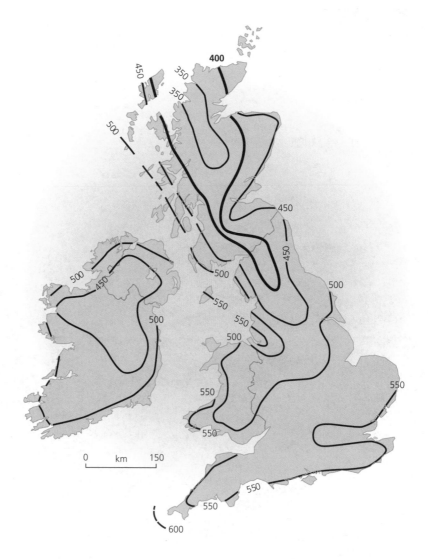

(b) The lowest amount of evapotranspiration is found towards the north of Scotland (less than 350 mm). The amount of evapotranspiration increases as you move south with high amounts found along the Welsh coast in the west and along the south coast of England (550 mm). The maximum is found on the tip of Cornwall in the southwest of the British Isles.

6 (a) 2

(b) 9

7 The graph is a bar chart.

It shows that 45 of the people surveyed came from Coleraine. This was the largest category, followed by Portrush (17), Ballymoney (13), Bushmills (11) and then Ballymena (10), Portstewart (9), Derry (5) and Ballycastle least with 4. The survey was conducted in Coleraine, the majority of the people surveyed

would be local. Generally, the further away that you travel from Coleraine, the fewer people came from that place.

8 The main areas of increase are Europe (+750,000) and North America (+1,000,000). The main areas of loss are Asia (–1,375,000) and Latin America & the Caribbean (–400,000).

9 Most of the squatter settlements are located in LEDCs, for example, in cities such as Mexico City, Lima, Rio and Brasilia in South America, Kinshasa and Dakar in Africa and Manila, Calcutta and Seoul in Asia.

10 (a) 25% hydroelectric, 52% nuclear 23% thermal and other.

(b) 10% hydroelectric, 48% nuclear and 42% thermal and other.

(c)

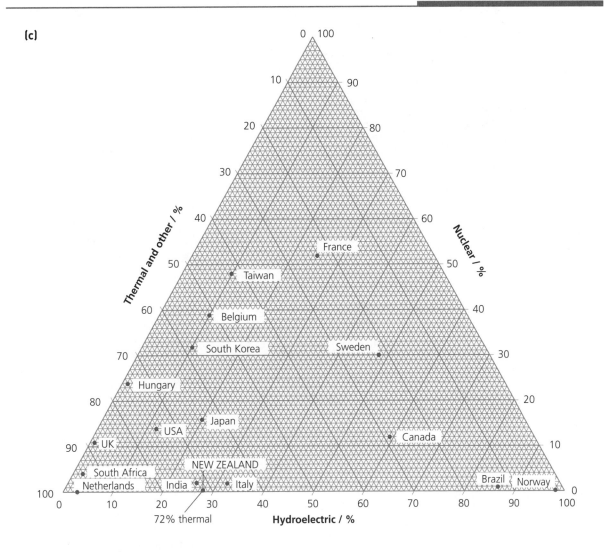

11 (a) Mean is 74, median is 74, mode is 76.

(b) The mean because it is the only measure that makes use of all of the values.

(c) The mode because it only records the value that occurs most times and this could be found at either extreme.

12 (a) The value is 99.9% significant.

(b) The Spearman's rank correlation shows that there is a strong negative relationship between the two variables. The reason for this is that in many countries it would be expected that, as the GNI PPP per capita (amount of wealth in the country) increases, so the number of births in the country would decrease. This might be due to a number of factors such as better access to contraceptive methods, better education and career prospects for women.

13 (a) Usually the minimum number of distances that needs to be measured is 30. In this case there are only 20 places/ distances and this might call into question the validity and accuracy of the result.

(b) If the map was focusing on one area, the R_n value would increase and the pattern would be more regular.

Index